i

Nuclear Rockets:

To the Moon and Mars

Manfred "Dutch" von Ehrenfried

Nuclear Rockets:
To the Moon and Mars

Front cover courtesy of NASA
Published with the help of Kindle Direct Publishing

Manfred "Dutch" von Ehrenfried
Cedar Park, TX. USA

Other Books by Manfred "Dutch" von Ehrenfried available on Amazon

Stratonauts: Pioneers Venturing into the Stratosphere, 2014
ISBN: 978-3-319-02900-9

The Birth of NASA: The Work of the Space Task Group, America's First True Space
Pioneers, 2016, ISBN: 978-3-319-28426-2

Exploring the Martian Moons: A Human Mission to Deimos and Phobos, 2017, ISBN:
978-3-319-52699-7

Apollo Mission Control: The Making of a National Historic Landmark, 2018, ISBN:
978-3-319-76683-6

From Cave Man to Cave Martian, Living in Caves on the
Earth, Moon and Mars, 2019. ISBN: 978-3-030-05407-6

The Artemis Lunar Program: Returning People to the Moon, 2020.
ISBN, 978-3-030-38512-5

Stratospheric Balloons: Science and Commerce at the Edge of Space, 2021, ISBN:
978-3-030-68129-6

Perseverance and the Mars 2020 Mission: Follow the Science to Jezero Crater, 2022,
ISBN: 978-3-030-92117-0

Nuclear Terrorism-A Primer, As it Applies to the U.S. Nuclear Industry, Second
Edition, 2022, ISBN: 9798847067034

The Search for Water on the Moon: Landers and Rovers in the 21st Century, 2022,
ISBN: 9798842720187

Asteroid Detection and Mitigation in the 21st Century, 2022,
ISBN: 9798844482571

Artemis Base Camp: The First Steps, 2022,
ISBN 9798359328784

Frontispiece

The rocket designs during the initial nuclear propulsion research efforts describe in this book were based on analog technologies. It was primarily after Apollo that the digital world began. Most would agree that the digital world is associated with the invention of the Internet and rise of home computers in 1969-1989. Over the past half century, the scientists and engineers have had time to revise their designs and give the artists an opportunity to visualize their concepts.

Nuclear Pulse Propulsion　　**Russian RD-0410**　　**NERVA NTP**
NASA 1958　　　　　　　　　**1960's**　　　　　**NASA 1964-69**

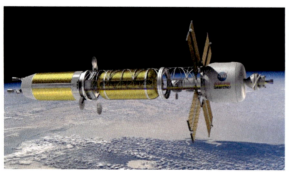
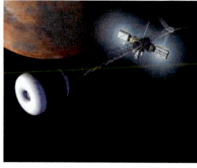

Nuclear Thermal Propulsion,　　**Nuclear Electric Propulsion**
NASA 1999　　　　　　　　　　**NASA 2004**

Dedication

It would be difficult to dedicate the book to all those who worked on nuclear rocket projects from those early days of the 1950's thru the 1970's. It would include tens of thousands of people. Most likely, they have all taken a spontaneous ride to elsewhere. While the funding for deep space travel research did not all dry up in 1973 when most of the work ended, it is hard to say just when projects that include nuclear propulsion started back up. Certainly, that type of research has benefitted from all the nuclear power research that has gone on. It has developed into a world-wide dependence on nuclear power for both civil and military use.

Nuclear researchers, physicists, engineers, technicians and administrators usually remain in the field and work on a host of projects. Traditionally, NASA, the DOD, and the DOE with their famous National Laboratories, are constant sources of funding and people for nuclear projects. Some of their work spins off to universities providing even more people and resources.

It appears that the modern movement in nuclear propulsion for deep space flight began in 2011 with the NASA Nuclear Cryogenic Propulsion Stage (NCPS) project. It has evolved over time as a result of a realignment of NASA's current priorities and missions. The effort is now part of a multi-agency alignment that includes many organizations within NASA, DOD, the National Laboratories, DARPA, universities and a dozen or more companies within the aerospace and nuclear industry. Our international colleagues in ESA, the UK, and the ISRO are also involved in this field of endeavor.

The nature of a goal such as sending humans to Mars and beyond is such that people may spend their entire careers supporting such an endeavor and then handing off to the next generation who may do the same. It is to all of these people; past, present and future, that this book is dedicated.

Acknowledgements

I would like to acknowledge all those people who have contributed to the new efforts and projects associated with developing the nuclear rockets needed to travel to Mars and other destination in deep space. Their bios and photos are in Appendix A3 Team Biographies, while many other scientists and engineers are mentioned in the Reference section, along with the titles of their reports.

Many thanks to the National Academies of Science for their analysis and reports on nuclear thermal propulsion, nuclear electric propulsion and nuclear surface power. Excerpts from these reports are included in this book.

In addition to contributions from many individuals, I acknowledge the help of Wikipedia and Google, which led me to other sources that allowed me to fill in the pieces of the puzzle on just about any subject. Their inputs are woven into many sections.

A special thanks to all the space journalists who have covered the ongoing activities over the past years. I have often read their reports and appreciate their contributions to spreading the knowledge of critical and timely events about this subject. I would also like to acknowledge those space journalists and enthusiasts who have video blogs on YouTube that add to the public interest in space exploration. Often, they are the first to report space related events and activities. Links to some of their videos are included.

I also thank NASA and ESA for their websites and inputs which were quite helpful. The various nuclear rocket projects are supported by many companies in the traditional nuclear and aerospace industries and are covered in the book. I found some of their websites very useful.

A special thanks to Dr. Dale Thomas of the University of Alabama at Huntsville for writing a Foreword as well as a special thanks to Dr. Tabitha Dodson, DARPA DRACO Program Manager, for writing a Foreword.

Attached are the names and affiliations of those administrators, scientists and engineers who are well known in the field; managing the various efforts concerning nuclear rockets and related projects. They come from NASA, DOD, DARPA, the DOE National Laboratories, supporting companies and universities. There are in alphabetical order within each category.

NASA HEADQUARTERS

Dr. Jacob Bleacher
Chief Exploration Scientist, HEOMD
NASA Headquarters
300 E Street SW
Washington, D.C., 20546

Dr. Anthony M. Calomino
Space Technology Mission Directorate (STMD).
NASA Headquarters
300 E Street SW
Washington, D.C., 20546

Col. (USAF, ret) Pam Melroy
Deputy Administrator
NASA Headquarters
300 E Street SW
Washington, D.C., 20546

James L. Reuter
Associate Administrator for STMD
NASA Headquarters
300 E Street SW
Washington, D.C., 20546

NASA CENTERS

Dr. Stanley K. Borowski
NASA John Glenn Research Center
21000 Brookpark Rd, Cleveland, OH 44135

Dr. William J. Emrich
NASA's Marshall Space Flight Center

Martin Rd SW, Huntsville, AL 35808

Marc Gibson
Space Technology Mission Directorate
NASA John Glenn Research Center
21000 Brookpark Rd, Cleveland, OH 44135

Dr. Michael G. Houts
Nuclear Research Manager
NASA's Marshall Space Flight Center
Martin Rd SW, Huntsville, AL 35808

Sonny Mitchell
Project Manager, Nuclear Thermal Propulsion
NASA's Marshall Space Flight Center
Martin Rd SW, Huntsville, AL 35808

Michelle Rucker
Mars Architecture Team
System Engineering and Integration Office
Kennedy Space Center, FL 32899

Dr. George Williams
Advanced Propulsion
NASA John Glenn Research Center
21000 Brookpark Rd, Cleveland, OH 44135

DARPA

Dr. Tabitha Dodson
DRACO Program Manager
675 North Randolph St.
Arlington, VA 22203

Dr. Stefanie Tompkins
Directory, DARPA
675 North Randolph St.
Arlington, VA 22203

DOE NATIONAL LABORATORIES

Dr. Stephen G. Johnson
Director of the Space Nuclear Power and Isotope Technologies Division
Idaho National Laboratory
1955 N Fremont Ave, Idaho Falls, ID 83415

Michael B. Smith
Advanced Reactor Engineering Group
Oak Ridge National Laboratory
5200, 1 Bethel Valley Rd, Oak Ridge, TN 37830

Dr. John C. Wagner
Director, Idaho National Laboratory
1955 N Fremont Ave, Idaho Falls, ID 83415

AEROSPACE INDUSTRY

Dr. Michael Eades
Chief Engineer, Tech Division
Ultra Safe Nuclear Corporation
2288 W Commodore Way Suite 300,
Seattle, WA 98199

David Durham
President, Energy Systems
Westinghouse Electric Company
Wilmington, NC

Dr. Pete Pappano
President, TRISO, LLC
Horizon Center Industrial Park,
Oak Ridge, TN 37830

Dr. Brad Rearden
Director of the Government R&D Division
X-energy
801 Thompson Ave Suite 400,
Rockville, Maryland, 20852

Dr. Paolo Venneri
CEO and founder of USNC-Tech
Ultra Safe Nuclear Corporation
2288 W Commodore Way Suite 300,
Seattle, WA 98199

Jonathan Witter
Chief Engineer Advance Technology Programs
800 Main Street
Lynchburg, VA 24504

UNIVERSITY

Dr. Dale Thomas
Deputy Director, Propulsion Research Center
University of Alabama Huntsville
320 Sparkman Drive,
Huntsville, AL 35899

Foreword by Dr. L. Dale Thomas

Fig. F.1 Dr. L. Dale Thomas

Space has fascinated humans for millennia. Since the beginning of humankind, the stars and planets in the sky have provided the backdrop for cultural myths and religions at the societal level and when to plant crops on a household level. Within recent generations, the word "spaceflight" emerged as people could now imagine actually traveling to some of those objects visible in the night sky. In 1865 the French author Jules Verne published the novel *De la Terre à la Lune* (From the Earth to the Moon) which involved landing three men on the Moon. More stories followed from other authors, stories more detailed in the means and methods of spaceflight and sometimes more outlandish as well. In the twentieth century engineers and scientists began taking the ideas and concepts for spaceflight from the realm of science fiction to the drawing board. The work of the engineers and scientists ultimately culminated in the United States' Apollo Program, which landed three astronauts on the Moon and safely returned them back to Earth, just over 100 years after Jules Verne's fanciful novel.

Following the Apollo Moon landings, mankind's passion for space travel appeared to cool. For the next fifty years, Earth's space faring nations were content to limit their space missions to low Earth orbits, as Russia continued to operate their 1960s era Soyuz rockets and the United States built the reusable Space Shuttle to construct and then to ferry their astronauts to and from the International Space Station. China joined the ranks of space faring nations as well, building and operating launch systems to fly their crews to and from low Earth orbit. But it is not only a game for the governments of the world's large industrialized nations; two private companies in the United States are now capable of spaceflight to and from low Earth orbit, as SpaceX and Boeing now ferry astronauts to and from the International Space Station. Spaceflight to and from Earth orbit has now become routine – a remarkable development in the short 150 years since Jules Verne first imagined spaceflight.

Those routine flights launching people to low Earth orbit are accomplished by vehicles unique to each government and private organization for reasons both strategic and pragmatic. But one thing that all the launch systems have in common is that they all rely on chemical combustion for propulsion. They all rely on rocket engines that mix a fuel – typically hydrogen or kerosene – with oxygen and use the very hot and high velocity exhaust stream from the rocket engine to propel the rocket. While these chemical powered rocket engines provide ample thrust for travel to the International Space Station or even to the Moon, missions within the Earth-Moon system are pretty much the limits of what can be achieved due to the chemistry and physics involved – rocket scientists sometimes refer to it as "the tyranny of the rocket equation." Mother Nature dictates the amount of energy released when two Hydrogen atoms are combined with a single Oxygen atom when combusted, and the generation of more thrust requires that more and more water molecules be formed. Today's rocket engines are remarkable technological achievements, using turbomachinery and intercoolers to efficiently combust the fuel and oxygen, and operate near the theoretical limits of efficiency. And this presents the prologue for this book – if mankind wishes to travel beyond the Earth-Moon system, more powerful rocket engines are needed, and significant improvements in chemical rocket engines are not possible.

In this book, Manfred "Dutch" von Ehrenfried introduces the propulsion technology which promises to make human spaceflight outside the Earth-Moon system possible – nuclear powered rockets. He begins with a review of prior work on nuclear propulsion. The United States actually developed and test fired (in the Nevada desert) Nuclear Rocket engines in the 1960s as the National Aeronautics and Space Administration prepared for the human missions to Mars that were planned to follow the Apollo Program's Lunar missions. However, the Nuclear Rocket engine program was scrapped when the United States chose to focus instead on low Earth orbit in the early 1970s. Within the last decade interest has rekindled in human missions to Mars, and the United States has again began investing in nuclear propulsion. Dutch describes the basics of the two primary approaches to nuclear propulsion – Nuclear Electric Propulsion (NEP) and Nuclear Thermal Propulsion (NTP) and discusses the advantages and drawbacks of each. He proceeds to describe the Demonstration Rocket for Agile Cislunar Operations (DRACO) project, a joint effort of the United States Defense Advanced Projects Agency and the National Aeronautics and Space Administration. The DRACO project plans to develop a NTP engine and fly a demonstration engine later this decade. This flight test will represent a big step toward a human mission to Mars in the 2030s.

Dutch has written this book for anyone who is curious about the possibilities of space travel and the potential of nuclear propulsion to enable humankind to travel to other planets for the first time. This book will help readers to understand the basics of nuclear propulsion, provide readers with a description of the research and development efforts currently underway, and inspire readers to ponder the possibilities and implications of nuclear propulsion. And perhaps one of the readers will be the first person to leave boot prints on the surface of Mars.

Ad Astra,
L. Dale Thomas, Ph.D., P.E.

Foreword by Dr. Tabitha Dodson

Fig. F.2 Phoebus 2A, shown prior to testing with engineers posing nearby. Photo courtesy of Threshold Magazine, 1992. This is a good illustration that shows how safe the nuclear thermal rockets are from a radiological perspective, if they are not turned on prior to launch. This approach would make launching nuclear thermal rockets simplified from the perspective of having negligible health hazards from a cold reactor in the event of a launch vehicle accident.

Technologies utilized by humans to cross vast distances have evolved over time and progressively allowed humans to break the bonds that have confined them to particular places on the Earth. From the means to create fire to migrate and stay warm in distant and cold climates, to the taming of horses to move and rebuild entire civilizations and habitats from one place to another. Those evolutions in technology and subsequent migrations that followed, took course over hundreds and thousands of years.

The invention of ocean-crossing ships, and the subsequent evolution in society's knowledge of the geometry of the Earth and non-existence of sea monsters, led to a surge in humans boarding ships and crossing the sea. Large air-breathing turbine engines and the perfection of airfoils allowed humankind to conquer and travel through the atmosphere, moving tons of people and cargo across continents. Finally, the invention of chemical rocket engines and the subsequent discovery that we could leave the surface of the Earth and push outside of its atmosphere has galvanized a following that seeks to go even farther, such that human-beings might be spacefaring.

Maneuvering and transportation require varying amounts of energy creation and application. From horse power (created naturally), to combustion power (chemical processes), to wind power, and finally to nuclear power – we have been able to understand, harness, and apply each means of energy generation to different modes of transportation. Nuclear energy is the most recently harnessed energy generating technique, and it is also the most energy dense with the most promise for pushing across farther distances. Submarines and large ships that incorporate nuclear reactors are the best and most successful example of how human beings have harnessed nuclear energy for transporting enormous ships weighing over 200,000 tons, across the distance of the entire planet.

A natural pairing occurs between humankind's greatest technological feat in generating vast amounts of nuclear energy, and for attempting humankind's most ambitious transportation feat: into outer-space. Nuclear technologies have been successfully utilized in the form of radioisotope thermoelectric generators (RTGs) for use by astronauts on the Moon during the Apollo program. RTGs have also been used in many missions for keeping deep-space platforms and Mars rovers operational when they are too far from the Sun to make use of solar energy.

Nuclear fission reactors offer an energy solution that can provide orders of magnitude more energy than the RTG. This increase in the amount of energy provided by the fission reactor is necessary, not only for providing the enormous amounts of electrical and thermal power needed to sustain human life in the inhospitable void of outer space, but also for pushing the very heavy cargo and platforms that will be crossing vast distances in outer space.

Generating usable energy from chemical combustion is a well understood process which is well-applied to transportation, using various automobile, ship, turbojet, and rocket engines. On the grand scale of time, humans just recently discovered (during the advent of the nuclear revolution the 1950's) that nuclear fission energy could readily replace combustion energy, with greater energy density and greater promise, for several different applications. One of those applications included rockets for space applications, as was envisioned during the Nuclear Engine for Rocket Vehicle Applications (NERVA) program which ran from 1953-1972.

Unfortunately, the propagation of knowledge for nuclear technologies had a shaky and difficult start in the mid-1900's, as culturally the association with nuclear technologies would sometimes invoke and be associated with nuclear weapons and the side-effects of nuclear radiation. However, over time, as the risks associated with nuclear technologies became well-understood, society has started to learn to control and be comfortable with nuclear technologies throughout the decades. Going into the early 2000's, as with any field, the more nuclear fission was understood, the less humankind had to fear from it.

An aggressive push by the global society to push into space, not just by the space agencies from the governments of the world, but also by the private sector for space transportation, is occurring in parallel by what has been called a "nuclear renaissance." Given the revived interest in nuclear technologies and understanding their promise, a well-spring of different design concepts and applications have multiplied and flourished. Naturally, a nexus has occurred where the push into space has merged with the growing and strengthening capabilities of the nuclear industrial base.

Nuclear propulsion falls broadly into two categories. Nuclear thermal rockets (NTRs) are more comparable to upper stage rocket engines that are in use today, in terms of their thrust and mission class. Nuclear electric propulsion (NEP) systems are defined as a nuclear power reactor that enables an electric thruster, such as a Hall thruster or ion thrusters. An NEP is not fast, like an NTR, but it can also push heavy platforms across vast distances to deep space.

There are uses for both NTR and NEP platforms, just as there continues to be uses for both upper stage rockets and deep space science missions. The performance enhancements that are provided by both of these propulsion systems open a new trade space of possibilities, when it comes to the types of missions that can be done in space. Current in-space propulsion systems (both electric and chemical) were sized and had their burn profiles tailored based on the understanding of Keplerian space at the time. Nuclear propulsion systems can now break the paradigm of propulsion systems confined to a 2D planar Kepler space, allowing mission designers to be creative within a spherical 3D volume, where simple orbital maneuvers can fractal into many possible directions of forced motion.

Breaking out of the simple 2D planar mission space may hopefully allow humans the ability to more responsively react to rapidly changing mission environments in real-time, in particular as space travel becomes less rote and automated, and more creative and adventurous. The evolution from chemical propulsion to NTR is akin to the evolution from driving cars on the road to now flying airplanes. Modeling tools and the approach to these new missions will similarly need to evolve, because just as it is imprudent to drive a plane on the ground like a car and use "driving tools" to model a planes performance on the ground, the tools and the mindset for planning missions enabled by NTR will need to evolve from 2D to 3D, as we take flight.

This is not the first time an up swell of interest in the United States has occurred to reinvigorate our nation's nuclear rocket propulsion capabilities of the past (i.e., Argonne's program, Timberwind, Prometheus, etc.). However, this time the up swell is occurring cross-agency, and cross-sector – most importantly – attracting and retaining university students who might actually have a job to look forward to and maintain continuity of their expertise after graduation. If the trend in the United States continues, then the "nuclear aerospace engineer" might just turn into its own specialty. And when our upcoming nation-wide nuclear space demonstrations, such as DRACO, are successful, they should provide the basis to land an entire workforce of hopeful young people into a future of meaningful and productive jobs, where nuclear aerospace hardware is built for real spaceflight missions.

Dr. Tabitha Dodson, PhD, DRACO Program Manager
Defense Advanced Research Projects Agency
Tactical Technology Office

Preface

Here we are, almost seven decades and at least two generations later, and we are still talking about nuclear rockets in space. Human spaceflight has seen many countries fly on chemical rockets to Earth orbit and a few to the Moon. As of today, people from 44 countries have traveled in space. About 622 people have reached Earth orbit, 628 have reached the altitude of space according to the FAI definition of the boundary of space, and 565 people have reached the altitude of space according to the American definition.

While we dream of flights to Mars, the farthest humans have been from Earth is just a little past the Moon; 400,171km (248,655 mi) and that was the Apollo 13 crew. That is just a "tippy toe" into deep space compared to the average distance to Mars at 225 million km (140 million mi). Obviously, we are still in our infancy as far as space travel is concerned.

With chemical rockets, the duration of round trip flights to Mars are typically listed as 545 days (18 months) for an Opposition Class (short stay) trajectory or a flight of 879 days (29 months) for a Conjunction Class (long stay) trajectory. These flight durations are almost intolerable; if not extremely risky. It is no wonder space planners have a renewed interest in nuclear rockets for deep spaceflights in order to cut the travel time down significantly. Wouldn't it be nice to cut that time in half?

A half century has passed since 1973 when the nascent nuclear rocket efforts ceased. Significant advancements in technology have benefitted the practicality of the use of nuclear rockets. These advancements are across the spectrum and include the movement from the analog world to the digital world. This change alone has certainly enhanced the designs and how the manufacturing world implements those designs with Computer Aided Design (CAD) and Computer Aided Manufacturing (CAM) systems, which were just invented during Apollo.

Not only have there been advances in the reactors; quantum leaps have been made in the design and manufacturing of the spacecraft systems and components. Coupled with the advancements in software development and lessons learned from deep space missions over the past half century; the state-of-the-art for nuclear thermal and electric propulsion rockets has significantly advanced and; it seems the world is ready for a trip to Mars. The advancements in launch vehicles and spacecraft were just demonstrated with the Artemis 1 mission in November of 2022.

Despite the technological advancements, deep space travels all boils down to the fuel. In this case, only hydrogen fuel since nuclear reactors don't need oxygen as there is no burning; no combustion. We see how much chemical fuel it took to get Artemis I to the Moon and back. Imagine how much chemical fuel and oxidizer it would take to go to Mars and back! Then consider the fact that uranium has seemingly unlimited energy available to heat only the hydrogen propellant. A typical manned Mars mission would require the energy released from the fissioning of only about 150 gm (0.333 lbs) of uranium! As one scientist put it; the nuclear rocket engines that NASA/DOD and aerospace industry are developing would only require a volume of uranium that is roughly the physical size of a marble. Obviously, the rocket would carry much more fuel but not having to carry an oxidizer leaves more mass (weight) allowance for cargo and humans.

Not only would the actual journey-time to deep space locations such as Mars be reduced, but the increased energy output would also allow for a wider launch window, including times when the Earth and Mars are not in the most optimal positions. Additionally, crewed missions on spacecraft with nuclear thermal propulsion engines would actually have the ability to abort their missions and return home before reaching their destination, which is an option that those missions would not have with chemically propelled engines due to their rapid propellant usage. Rather than refueling at their destination before turning home, crews would be able to make that decision mid-transit.

The most critical element in the design of advanced nuclear reactors is a robust fuel that can withstand extremely high temperatures without melting caused by a propellant (hydrogen) reactor exit temperature of approximately 2700° K (4400 ° F). This represents an extreme environment in terms of temperature and hydrogen corrosion for the materials in the reactor core. This extreme reactor operating temperature implies that there are few viable fuel architectures. The fuel element, which includes the fuel and cladding, the fuel assemblies, moderator, support structures, and the reactor pressure vessel, must maintain physical integrity while cycled through the thermomechanical stress induced during repeated cycles of reactor startup, operation at power, shutdown, and restart.

The scientific and engineering advances have even impacted the world of nuclear fuels including the switch to High-Assay Low-Enriched (HALEU) uranium. While a relatively new and unique fuel called, Tri-Structural Isotropic (TRISO) fuel was first developed in the U.S. and U.K. in the 1960s with uranium dioxide fuel; significant advancements have been made since 1973.

In 2002, the DOE focused on improving TRISO fuel using uranium oxycarbide fuel kernels. In 2009, this improved TRISO fuel set an international burnup test record. The amount of heat that can be generated from these new fuels is significant enough to accelerate gaseous hydrogen to unbelievable velocities; enough to double (or more) the efficiency of a chemical rocket; the result being, to cut the time to and from Mars in half. Needless to say, this lowers the risk to the crew and increases the probability of mission success.

But alas; there is a more pressing commitment that must be addressed before a trip to Mars; the return to the Moon and the establishment of a permanent human presence. In the meantime, the evolution of the nuclear thermal and electric rockets will have to proceed thru their interminably slow development and test phases. At least there are now scientists and engineers focused on the goal of getting to Mars faster, with fewer risks but with more flight opportunities.

The reader might consider this as a reference book available over the long period of time that it will take to develop nuclear rockets for travel to the Moon and Mars.

Manfred "Dutch" von Ehrenfried
Cedar Park, TX
Passing the Vernal Equinox, 2023

1

Introduction

It is important at the outset, to make the distinction between nuclear power for making heat to be turned into electricity and nuclear power for propulsion. And of course, this book has nothing to do with making nuclear bombs. For over 60 years, nuclear power, in one form or another, has been used in space; including todays Martian explorers including Curiosity and Perseverance to name a few.

Typically, the use of nuclear power in space is either small fission systems that use radioactive decay for electricity or heat. The most common type is a radioisotope thermoelectric generator (RTG), which has been used on many space probes such as Viking, Voyager and Pioneer as well as for the Apollo Lunar Surface Experiment Packages (ALSEP). Small fission reactors for Earth observation satellites have also been flown. A radioisotope heater unit is powered by radioactive decay and can keep components from becoming too cold to function; potentially over a span of decades.

This book focuses on propulsion but also addresses nuclear fission power to provide electricity. Examples of concepts that use nuclear power for space propulsion systems include the nuclear electric rocket (nuclear powered ion thruster), the radioisotope rocket, and radioisotope electric propulsion (REP). One of the more explored concepts is the nuclear thermal rocket, which was ground tested during the NERVA program and explained in Chapter 2.

For an 8:53 minute video on the various concepts of propulsion, go to:
https://www.youtube.com/watch?v=p0xfVJ1WTf4

For a 3:04 minute introductory video on the Nuclear Thermal Propulsion concept, go to:
https://www.youtube.com/watch?feature=youtu.be&v=miy2mbs2zAQ&app=desktop&ab_channel=NASAVideo

So why is NASA refocusing its efforts on Nuclear Thermal Propulsion now, when chemical propulsion is so well established for human spaceflight? The reason is that future proposed flights are getting much longer and studies have shown that long-duration, weightless spaceflight has a lot of deleterious effects on the human body. When considering a flight to Mars takes a minimum of 2-3 years roundtrip, anything we can do to shorten the total mission time is highly desirable; if not mandatory. The longer the mission is, the more exposure the systems and crew to galactic cosmic radiation, solar particles and coronal mass ejections, potential medical problems and emergencies.

While only a few astronauts and cosmonauts have spent a continuous year in space, Valeri Vladimirovich Polyakov (born 4/27/1942- died 9/7/2022 at age 80) is the record holder for the longest single stay in space, staying aboard the Mir space station for more than 14 months (437 days 18 hours) during one trip in1994 and 1995.

Polyakov experienced a clear decline in mood as well as a feeling of increased workload during the first few weeks of spaceflight and return to Earth. Polyakov's mood stabilized to pre-flight levels between the second and fourteenth month of his mission. It was also revealed that Polyakov did not suffer from any prolonged performance impairments after returning to Earth. In light of these findings, researchers concluded that a stable mood and overall function could be maintained during extended duration spaceflights, such as crewed missions to Mars. However, it was clear that more mitigation of the effects of long duration spaceflight was needed.

Fast forward three decades; after a generation of orbital spaceflight, a little over 600 people have been to space, but no one has been continually in space for the equivalent time of a Mars mission. There are several others that have accumulated more total time in space than Polyakov, but over several missions and years. Cosmonaut Gennady Padalka holds the record for most time spent in space over multiple missions. Over the course of seventeen years and five space flights, Padalka spent a total of 878-days in space. This equates to nearly two and a half years in orbit, but does not equate to the same time in deep space with its increased radiation, the stress of a Mars trip and fewer opportunities to return to Earth.

In recent years, the objectives of the space program have grown increasingly sophisticated and ambitious. Future missions will focus on exploration at greater distances from Earth and extended stays in space. To ensure that these goals are achieved, NASA's astronauts must be able to perform at peak productivity under even the most daunting conditions. The NASA Human Research Program (HRP) is dedicated to discovering the best methods and technologies to support safe, productive human space travel.

From the challenges of providing appetizing food and optimal nutrition to managing the environmental risks posed by radiation, HRP scientists and engineers work to predict, assess, and solve the problems that humans encounter in space. Planned future missions will dramatically increase the scope of the challenges and demands that face NASA's astronauts. The HRP is working to improve astronauts' ability to collect data, solve problems, respond to emergencies, and remain healthy during and after extended space travel. Certainly, reducing the duration of exposure to the dangers of deep space is high on the list.

As a result of NASA's refocus of the space program on exploration, NASA uses research findings to develop procedures to lessen the effects of the space environment on the health and performance of humans working in deep space for extended periods of time. With the goal of traveling to Mars and beyond, the HRP is using ground research facilities, the International Space Station, and analog environments to develop these procedures and to further research areas that are unique to Mars.

The Human Research Program is comprised of five Elements; they are:

1. International Space Station Medical Projects,
2. Space Radiation,
3. Human Health Countermeasures,
4. Exploration Medical Capability and,
5. Human Factors and Behavioral Performance.

The HRP is dedicated to discovering the best methods and technologies to support safe, productive human space travel. It is working to improve the astronauts' ability to collect data, solve problems, respond to emergencies, and remain healthy during and after extended space travel. The program investigators work to predict, assess, and solve the problems that humans encounter in space. It leverages assets through national and international collaborations, and educates and engages the public about the challenges of human space travel.

A first-generation nuclear cryogenic propulsion system could propel human explorers to Mars more efficiently than conventional spacecraft, reducing the crews' exposure to harmful space radiation and other effects of long-term space missions. It could also transport heavy cargo and science payloads.

Maturing faster, more efficient transportation technology will help NASA meet its Moon to Mars objectives. In the late 50's and early 60's, NASA worked on the nuclear thermal propulsion systems, but ceased testing when plans for a crewed Mars mission were deferred. But now, generations later, NASA has turned its focus on going back to the Moon with the intent to take what is learned from the Artemis program and apply the lessons learned to the Mars program.

Nuclear Thermal Propulsion (NTP) can help enable detailed exploration of the solar system, be it enhancing operations in cis-lunar space, shaping robust human Mars mission plans or a variety of outer planet space science endeavors.

A nuclear rocket engine uses a nuclear reactor to heat hydrogen to very high temperatures, which expands through a nozzle to generate thrust. Nuclear rocket engines generate higher thrust and are more than twice as efficient as conventional chemical rocket engines; just what's needed for the crew to better survive the long duration missions to Mars.

NTP engines will be able to complete crewed Mars missions significantly faster than chemical propulsion. While traditional chemical engines are used for launch, the NTP rockets are only used in space. But should an explosion occur during the launch phase, having a NTP upper stage does not pose an excess risk. The core is designed to withstand water submersion without activation. This means explosive launch failures or water submersion scenarios do not lead to nuclear accidents and pose no additional risk compared to ordinary launch failures.

Chapter 2 provides a discussion of the first attempts to study and test nuclear thermal propulsion. It describes the Air Force efforts in the 1950's up to when NASA was formed and took over the projects. They included the Rover Program with its Kiwi, Phoebus and NERVA projects until they were cancelled in 1973 due to costs of the Vietnam War and the Apollo program which chose chemical rockets.

Chapter 3 describes the governments' efforts including DARPA's DRACO program and the NASA Centers support efforts including the Glenn Research Center, Marshall Space Flight Center and the Stennis Space Center. It also includes the DOE's National Laboratories efforts in Nuclear Thermal Propulsion research including the Idaho National Laboratory, the Oak Ridge National Laboratory and the Los Alamos National Laboratory.

Chapter 4 describes all the aerospace industry contractors' efforts to support the government programs involving nuclear thermal propulsion, nuclear electric propulsion and nuclear fission surface power. Four major aerospace teams are described for nuclear thermal propulsion that includes ten companies. These contracts are, for the most part over; awaiting analysis and possibly new contracts.

Chapter 5 describes the technologies associated with nuclear fission surface power for the Moon and Mars and the three teams consisting of nine companies working on these projects. They are in the same posture as the contractors above.

Chapter 6 describes the Baseline Considerations that drive the technologies needed for deep space transportation, and exploration. This chapter discusses the new nuclear fuels, how they are made, the unique micro reactors, support systems and the difficulties of cryogenic storage of hydrogen over long periods in deep space. It describes moderators, reflectors, and heat pipes of various types.

Chapter 7 describes the National Academies of Science report on Space Nuclear Propulsion for Human Mars Exploration. Based on their extensive team's report, they list their "Findings and Recommendations" of what needs to be addressed for both Nuclear Thermal Propulsion and Nuclear Electric Propulsion.

Chapter 8 describes the activities of international organizational efforts in nuclear thermal and electrical propulsion including the ESA, UK, Roscosmos, China and India. Surprisingly, a commercial company, Rolls-Royce is also involved in nuclear reactor design.

Chapter 9 Conclusions. The author gives his views on where the development of nuclear rockets has been and what the future holds for their use in deep space flight. Points are made about the advancement of technology since the age of the early nuclear rocket program and how the Internet and the digital world have given us immediate access to information. Technology advancements have been made in many areas, yet some are very old but proven to apply to new applications

Major efforts are now underway within NASA, the DOD and DOE as well as with their selected contractors, to meet the challenges laid out by deep space exploration. So too, are efforts underway with our International partners and our competitors.

While those challenges are partially being met by the application of nuclear propulsion, it is clear in hindsight, that we definitely did not have the capability to go to Mars in those early days; we barely have the capability today. The goal to get humans to Mars is now more than a decade away; probably longer.

2

The Beginning of the Nuclear Rocket Program

2.1 ROVER

2.1.1 The Concept

After World War II, engineers became interested in utilizing the massive power of atomic fission for aircraft and missile propulsion. In 1945 the military began sponsoring efforts to develop an atomic aircraft. Engineers, however, could not overcome issues involving the required shielding for the crew or the fear of radiation at crash sites. Selecting the fuels to use is a complex task; they had to consider the reaction energy, the mass of the fuel, the mass of the resulting working fluid, and other practical concerns like density and its ability to be easily pumped.

Rocket engines create thrust by accelerating a working mass in a direction opposite to their desired trajectory. In conventional designs, this is accomplished by heating a fluid and allowing it to escape through a rocket nozzle. The energy needed to produce the heat is provided by a chemical reaction in the fuel; typically mixing and burning a fuel with an oxidizer such as hydrogen and oxygen as on the Space Shuttle and Artemis launch vehicles. Often the fuel and oxidizer is mixed together as in the case of most solid fuel rockets, such as those on the sides of those vehicles. But nuclear thermal propulsion rockets are not designed for launch vehicles; they are intended for deep space propulsion.

Nuclear rocket engines use a nuclear reactor to provide the energy to heat the fuel instead of a chemical reaction. They don't need an oxidizer to burn the fuel; they just need the heat from the reactor. Because nuclear reactions are much more powerful than chemical ones, a large volume of chemicals can be replaced by a small reactor. As the heat source is independent of the working mass, the working fluid can be selected for maximum performance for a given task, not its underlying reaction energy. For a variety of reasons, hydrogen is normally used. These features allow a nuclear engine to outperform a chemical one; having at least twice the specific impulse of a chemical engine. In general form, a nuclear engine is similar to a liquid chemical engine. Both hold the working mass in a large tank(s) and pump it to the reaction chamber using a turbopumps. Robert Goddard was aware of these complicating factors even in the 1920s. Wernher von Braun perfected them in the 1940's;

> 1. The first was that a means had to be found of controlling reactor temperature and power output.
> 2. The second was that a means had to be devised to hold the propellant. The only practical means of storing hydrogen was in liquid form, and this required temperatures below 20 K (−253.2 °C).
> 3. The third was that the hydrogen would be heated to a temperature of around 2,500 K (2,230 °C), and materials were required that could both withstand such temperatures and resist corrosion by hydrogen.

For the nuclear rocket fuel, plutonium-239, uranium-235 and uranium-233 were considered. Plutonium was rejected because it forms compounds easily and could not reach temperatures as high as those of uranium. Uranium-233, as compared to uranium-235, is slightly lighter, has a higher number of neutrons per fission event, and also has high probability of fission, but its radioactive properties make it more difficult to handle, and it was not readily available. Uranium-235 was therefore chosen.

For structural materials in the reactor, the choice came down to graphite or metal. Of the metals, tungsten emerged as the frontrunner, but it was expensive, hard to fabricate, and had undesirable neutronic properties. To get around these properties, it was suggested tungsten-184, which does not absorb neutrons, should be used. On the other hand, graphite was cheap, actually gets stronger at temperatures up to 3,300 K (3,030 °C), and sublimes rather than melts at 3,900 K (3,630 °C). Graphite was therefore chosen.

To control the reactor, the core was surrounded by control drums coated with graphite or beryllium (a neutron moderator) on one side and boron (a neutron poison) on the other. The reactor's power output could be controlled by rotating the drums. To increase thrust, it is sufficient to increase the flow of propellant. Hydrogen, whether in pure form or in a compound like ammonia, is an efficient nuclear moderator; increasing the flow also increases the rate of reactions in the core. This increased reaction rate offsets the cooling provided by the hydrogen. Moreover, as the hydrogen heats up, it expands, so there is less in the core to remove heat, and the temperature will level off. These opposing effects stabilize the reactivity and a nuclear rocket engine is therefore naturally very stable, and the thrust is easily controlled by varying the hydrogen flow without changing the control drums.

The Los Alamos Scientific Laboratory (LASL) now the Los Alamos National Laboratory (LANL) produced a series of design concepts and in 1955, settled on a 1,500 MW design. In 1956, this became the basis of a 2,700 MW design intended to be the upper stage of an ICBM.

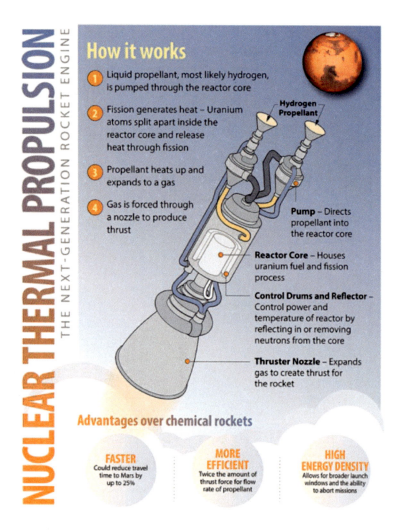

Fig. 2.1 The simplicity of a Nuclear Thermal Propulsion Rocket. Illustration courtesy of U.S. DOD Office of Nuclear Energy.

2.1.2 From Concept to Project

In 1955 the military partnered with the Atomic Energy Commission (AEC) to develop reactors for nuclear rockets under Project Rover, an Air Force project to develop a nuclear-thermal rocket that ran from 1955 to 1973 at the. Its purpose was to develop a nuclear-powered upper stage for an intercontinental ballistic missile. The upper stage would not be fired until it entered space reducing the threat of crash-induced contamination on Earth.

Shortly after NASA was formed in 1958, the Marshall Space Flight Center was created with rocket pioneer Wernher von Braun as its Director. His vision was to use Nuclear Thermal Propulsion (NTP) to send astronauts to Mars by the early 1980s. He proposed dispatching a dozen crew members to Mars aboard two rockets. Each rocket would be propelled by three NERVA engines. As detailed by von Braun, that expeditionary crew would launch to the Red Planet in November 1981 and land on that distant world in August 1982. In presenting his visionary plan in August 1969 to a Space Task Group, von Braun explained that "although the undertaking of this mission will be a great national challenge, it represents no greater challenge than the commitment made in 1961 to land a man on the moon." It seems his vision was very ambitious. Current estimates for crewed missions to Mars using chemical rockets, begins in the 2030's. Missions using NTP would most likely be even later. His vision for a heavy lift launch vehicle for the Moon was right on. On Nov. 9, 1967, the first Saturn V launched.

2.1.3 Kiwi Reactors

In 1959, NASA replaced the Air Force in the development of the rocket and the mission changed from a nuclear missile to a nuclear rocket for long-duration space flight. The Rover program began with research on basic reactor and fuel systems. This was followed by a series of Kiwi reactors built to test nuclear rocket principles in a non-flying nuclear engine. The first phase of Project Rover, Kiwi, was named after the flightless bird of the same name from New Zealand, as the Kiwi rocket engines were not intended to fly either. Their function was to verify the design and test the behavior of the materials used. The Kiwi program developed a series of non-flyable test nuclear engines, with the primary focus on improving the technology of hydrogen-cooled reactors. Between 1959 and 1964, a total of eight reactors were built and tested. Kiwi was considered to have served as a proof of concept for nuclear rocket engines.

Fig. 2.1 Kiwi-A Prime at the Nevada Test Site in Jackass Flats, NV. Photo courtesy of AEC/LANL.

The above photo shows the preparation of a Kiwi-A reactor for a test at Los Alamos Scientific Laboratory on 11/30/1959.

Fig. 2.3 Kiwi B-1. Photo courtesy of NASA/GRC.

The above photo shows technicians in a vacuum furnace at the Lewis Fabrication Shop preparing a Kiwi B-1 nozzle for testing in the B-1 test stand on 5/8/1964.

For a 60 year old, 20 minute film on Project Rover, go to:
https://www.youtube.com/watch?v=866C4qKgzeg

For an 18:04 minute film from the 1960's on how a Nuclear Rocket works, go to:https://www.youtube.com/watch?v=Zm7PNlK5Aco&list=RDLVU1g2aSj9ZTc&start_radio=1

Aerojet was simultaneously incorporating one of the Kiwi-B reactor designs into its NERVA NRX (NERVA Reactor Experiment) engine.

2.1.4 Phoebus

Phoebus was a series of nuclear reactors, designed and built in the 1960s as part of the Rover program, to meet the needs of an interplanetary mission, in particular a manned mission to Mars. The design requirements were a thrust of 250,000 lb, a specific impulse (Isp) of 840 sec, and a reactor power level of 5,000 MWt.

The Phoebus-1 series was intended to study increasing the reactor power density and proved successful. When the power density was increased still further in the Phoebus-2 series, however, cooling of the aluminum pressure vessel was found to be a limiting factor. The Phoebus series demonstrated:

> 1) basic core and fuel configuration technology,
> 2) control of rocket parameters over a wide range of operating conditions,
> 3) the niobium carbide-molybdenum (NbC-Mo) coating could protect the fuel elements from hydrogen corrosion,
> 4) a two-pass regeneratively-cooled support structure allows full core performance and,
> 5) large nozzles for nuclear thermal rocket application were feasible.

Other reactors developed during the Rover program were KIWI, Pewee-1, and Nuclear Furnace 1.

Fig. 2.4 Phoebus Nuclear Rocket on the Jackass & Western railroad.
Photo courtesy of the DOE.

The next phase, the Nuclear Engine for Rocket Vehicle Application (NERVA), sought to develop a flyable engine. The final phase of the program, called Reactor-In-Flight-Test, would be an actual launch test.

The AEC worked to develop the reactor for the engine at its facilities in New Mexico and Nevada, and the Lewis Research Center (now the Glenn Research Center) concentrated its efforts on the vehicle's liquid-hydrogen system. The Rocket Systems Area provided resources to conduct basic research on nuclear engine systems and to test hydrogen pumping systems. A series of 300-megawatt Kiwi-A reactors were tested at the Nevada Test Site in 1959 and 1960. The Kiwi-B reactors, which dramatically increased the power without increasing the overall size, were tested between 1961 and 1964.

2.1.5 Pewee

Pewee was the third phase of Project Rover. LASL reverted to bird names, naming it after the North American Pewee. It was small, easy to test and a convenient size for uncrewed scientific interplanetary missions or small nuclear "tugs". Its main purpose was to test advanced fuel elements without the expense of a full-sized engine. Pewee took only nineteen months to develop from when SNPO authorized it in June 1967 to its first full-scale test in December 1968.

Pewee had a 53 cm (21 in) core containing 36 kg (80 lb), 402 fuel elements and 132 support elements. Of the 402 fuel elements, 267 were fabricated by LASL, 124 by the Westinghouse Astronuclear Laboratory, and 11 at the AEC's Y-12 National Security Complex. Most were coated with niobium carbide (NbC) but some were coated with zirconium carbide (ZrC) instead; most also had a protective molybdenum coating. There were concerns that a reactor so small might not achieve criticality, so zirconium hydride (a good moderator) was added, and the thickness of the beryllium reflector was increased to 20 cm (8 in). There were nine control drums. The whole reactor, including the aluminum pressure vessel, weighed 2,570 kg (5,670 lb).

Pewee 1 was started up three times: for check out on November 15, 1968, for a short duration test on November 21, and for full power endurance test on December 4, 1968. The full power test had two holds during which the reactor was run at 503 MW (1.2 MW per fuel element). The average exit gas temperature was 2,550 K (2,280 °C); the highest ever recorded by Project Rover. The chamber temperature was 2,750 K (2,480 °C), another record. The test showed that the zircon carbide was more effective at preventing corrosion than niobium carbide. No particular effort had been made to maximize the specific impulse, that not being the reactor's purpose, but Pewee achieved a vacuum specific impulse of 901 seconds (8.84 km/s), well above the target for NERVA. So too was the average power density of 2,340 MW/m3; the peak density reached 5,200 MW/m3. This was 20% higher than Phoebus 2A, and the conclusion was that it might be possible to build a lighter yet more powerful engine still.

LASL took a year to modify the Pewee design to solve the problem of overheating. In 1970, Pewee 2 was readied in Test Cell C for a series of tests.

LASL planned to do twelve full-power runs at 2,427 K (2,154 °C), each lasting for ten minutes, with a cooldown to 540 K (267 °C) between each test. SNPO ordered LASL to return Pewee to E-MAD. The problem was the National Environmental Policy Act (NEPA), which President Richard Nixon had signed into law on January1, 1970. SNPO believed that radioactive emissions were well within the guidelines, and would have no adverse environmental effects, but an environmental group claimed otherwise. SNPO prepared a full environmental impact study for the upcoming Nuclear Furnace tests. In the meantime, LASL planned a Pewee 3 test. This would be tested horizontally, with a scrubber to remove fission products from the exhaust plume. It also planned a Pewee 4 to test fuels, and a Pewee 5 to test afterburners. None of these tests were ever carried out.

2.1.6 Nuclear Furnace

Two of the fuel forms tested by Project Rover: pyrolytic carbon-coated uranium carbide fuel particles dispersed in a graphite substrate, and a "composite" which consisted of uranium carbide-zirconium carbide dispersion in the graphite substrate.

The Nuclear Furnace was a small reactor only a tenth of the size of Pewee that was intended to provide an inexpensive means of conducting tests. Originally it was to be used at Los Alamos, but the cost of creating a suitable test site was greater than that of using Test Cell C. It had a tiny core 146 cm (57 in) long and 34 cm (13 in) in diameter that held 49 hexagonal fuel elements. Of these, 47 were uranium carbide-zirconium carbide "composite" fuel cells and two contained a seven-element cluster of single-hole pure uranium-zirconium carbide fuel cells. Neither type had previously been tested in a nuclear rocket propulsion reactor. In all, this was about 5 kg of highly enriched (93%) uranium-235. To achieve criticality with so little fuel, the beryllium reflector was over 36 cm (14 in) thick. Each fuel cell had its own cooling and moderating water jacket. Gaseous hydrogen was used instead of liquid to save money. A scrubber was developed.

The objectives of the Nuclear Furnace tests were to verify the design, and test the new composite fuels. Between June 29, and July 27, 1972, NF-1 was operated four times at full power (44 MW) and a fuel exit gas temperature of

2,444 K (2,171 °C) for a total of 108.8 minutes. The NF-1 was operated 121.1 minutes with a fuel exit gas temperature above 2,222 K (1,949 °C). It also achieved an average power density 4,500 to 5,000 MW/m3 with temperatures up to 2,500 K (2,230 °C). The scrubber worked well, although some krypton-85 leaked. The Environmental Protection Agency was able to detect minute amounts, but none outside the test range.

The tests indicated that composite fuel cells would be good for two to six hours operation at 2,500 to 2,800 K (2,230 to 2,530 °C), which the carbide fuels would give similar performance at 3,000 to 3,200 K (2,730 to 2,930 °C), assuming that problems with cracking could be overcome with improved design. For ten hours of operation, graphite-matrix would be limited to 2,200 to 2,300 K (1,930 to 2,030 °C); the composite could go up to 2,480 K (2,210 °C), and the pure carbide to 3,000 K (2,730 °C). Thus, the test program ended with three viable forms of fuel cell.

2.1.7 Safety tests

In May 1961, President Kennedy gave his approval for reactor in-flight tests (RIFT). In response, LASL established a Rover Flight Safety Office, and SNPO created a Rover Flight Safety Panel, which supported RIFT. NASA's RIFT planning called for up to four reactors to fall into the Atlantic Ocean. LASL had to determine what would happen when a reactor hit the water at several thousand kilometers per hour. In particular, it needed to know whether it would go critical or explode when flooded with seawater; a neutron moderator. There was also concern about what would happen when it sank 3.2 km (2 mi) down to the bottom of the Atlantic, where it would be under a crushing pressure. The possible impact on marine life, and indeed what marine life was down there, all had to be considered.

A modified Kiwi nuclear reactor was deliberately destroyed in the Kiwi TNT test. LASL started by immersing fuel elements in water. It then went on to conduct a simulated water entry test (SWET) during which a 30-cm (12 in) piston was used to force water into a reactor as fast as possible. To simulate an impact, a mock reactor was dropped onto concrete from a height of 23 m (75 ft). It bounced 4.6 m (15 ft) in the air; the pressure vessel was dented and many fuel elements were cracked but calculations showed that it would neither go critical nor explode. However, RIFT involved NERVA sitting atop a Saturn V rocket 91 m (300 ft) high. To find out what would happen if the booster exploded on the launch pad, a mock reactor was slammed into a concrete wall using a rocket sled. The core was compressed by 5%, and calculations showed that the core would indeed go critical and explode, with a force equivalent to about 2 kg (4.4 lb) of high explosive, which would likely be negligible compared to the damage caused by an exploding booster. Disturbingly, this was much lower than the 11 kg (25 lb) that was predicted theoretically, indicating that the mathematical modeling was deficient.

When it was determined that NERVA was not required for Apollo, and would therefore not be needed until the 1970s, RIFT was postponed, and then canceled entirely in December 1963. Although its reinstatement was frequently discussed, it never occurred. This eliminated the need for further SWET, but concerns remained about the safety of nuclear rocket engines. While an impact or an explosion could not cause a nuclear explosion, LASL was concerned about what would happen if the reactor overheated. A test was devised to create the most devastating catastrophe possible. A special test was devised known as Kiwi-TNT. Normally the control drums rotated at a maximum speed of 45° per second to the fully open position at 180°. This was too slow for the devastating explosion sought, so for Kiwi-TNT they were modified to rotate at 4,000° per second. The test was carried out on January 12, 1965. Kiwi-TNT was mounted on a flatbed railroad car, nicknamed the Toonerville Trolley, and parked 190 m (630 ft) from Test Cell C. The drums were rotated to the maximum setting at 4,000° per second and the heat vaporized some of the graphite, resulting in a colorful explosion that sent fuel elements flying through the air, followed by a highly radioactive cloud with radioactivity estimated at 1.6 megacuries (59 PBq).

Most of the radioactivity in the cloud was in the form of caesium-138, strontium-92, iodine-134, zirconium-97 and krypton-88, which have short half-lives measured in minutes or hours. The cloud rose 790 m (2,600 ft) into the air and drifted southwest, eventually blowing over Los Angeles and out to sea. It was tracked by two Public Health Service (PHS) aircraft which took samples. The PHS had issued film badge dosimeters to people living on the edge of the test area, and took milk samples from dairy farms in the cloud's path. They revealed that exposure to people living outside the Nevada Test Site was negligible. Radioactive fallout on the ground also dissipated rapidly. Search teams scoured the area collecting debris. The largest was a piece of the pressure vessel weighing 67 kg (148 lb) which was found 230 m (750 ft) away; another, weighing 44 kg (98 lb) was found 520 m (1,700 ft) away.

The explosion was relatively small, estimated as being the equivalent of 90 to 140 kg (200 to 300 lb) of black powder. It was far less violent than an explosion of TNT, and hence the large pieces that were found. The test showed that the reactor could not be destroyed in space by blowing it up into small pieces, so another method had to be found for disposing of it at the end of a space mission. LASL decided to take advantage of the engine's restartability to dispose of a nuclear rocket by firing it into a high orbit, so it either left the Solar System entirely or returned centuries later, by which time most of the radioactivity would have decayed away. The Soviet Union protested the test, claiming that it was a nuclear test in violation of the Partial Nuclear Test Ban Treaty, but the US replied that it was a subcritical test involving no explosion. However, the State Department was very unhappy with LASL's Kiwi-TNT designation, as this implied an explosion, and it made it harder to charge the Soviets with violating the treaty.

There were three fatal accidents during Project Rover. One worker was killed in a motor vehicle accident. Another died from burns after tipping gasoline on classified computer tapes and setting them alight to dispose of them. A third entered a nitrogen tank and was asphyxiated.

2.1.8 Cancellation

Rover was always a controversial project, and defending it from critics required a series of bureaucratic and political battles. In the late 1960s, the rising cost of the Vietnam War put increased pressure on budgets. The cost-cutting pressure increased after Nixon replaced Johnson as President in 1969. Over the years, Congress support for the nuclear thermal propulsion projects including the Saturn upper stage, lunar and Mars missions and the "Grand Tour" of the Solar System waned. On January 5, 1973, NASA announced that NERVA (and therefore Rover) was terminated. After 17 years of research and development, Projects Rover and NERVA had spent about $1.4 billion, but no nuclear-powered rocket has ever flown.

2.2 NERVA

The Nuclear Engine for Rocket Vehicle Application (NERVA) had its origins in Project Rover. Nuclear rocket engines promised to be more efficient than chemical ones. After the formation of NASA in 1958, Project Rover was continued as a civilian project and was reoriented to producing a nuclear powered upper stage for NASA's Saturn V Moon rocket. Reactors were tested at very low power before being shipped to Jackass Flats at the Nevada Test Site. While LASL concentrated on reactor development, NASA built and tested complete rocket engines.

The NERVA was a nuclear thermal rocket engine development program that ran for almost two decades. Its principal objective was to "establish a technology base for nuclear rocket engine systems to be utilized in the design and development of propulsion systems for space mission application". It was managed by the Marshal Space Flight Center's (MSFC) Space Nuclear Propulsion Office (SNPO) until the program ended in January 1973.

The first NERVA NRX test was run in September 1964 in Nevada. In 1969, the AEC successfully tested a second-generation NERVA engine, the XE, dozens of times. Funding for NERVA, however, decreased in the late 1960s and the program was cancelled in 1973 before any flight tests of the engine took place.

The AEC, SNPO, and NASA considered NERVA to be a highly successful program in that it met or exceeded its program goals. NERVA demonstrated that nuclear thermal rocket engines were a feasible and reliable tool for space exploration and, at the end of 1968; SNPO deemed that the latest NERVA engine, the XE, met the requirements for a human mission to Mars. Although NERVA engines were built and tested as much as possible with flight-certified components and the engine was deemed ready for integration into a spacecraft; they never flew in space.

Fig. 2.5 NERVA XE Nuclear Thermal Rocket. Photo courtesy of the National Nuclear Security Administration Nevada Site Office.

The above photo was taken at the Experimental Test Stand-1 at Jackass Flats, Nevada.

Specifications

Thrust, vacuum   246,663 N (55,452 lbf)

Chamber pressure	3,861 kPa (560 psi)
Specific impulse, vacuum	841 seconds (8.25 km/s)
Specific impulse, sea-level	710 seconds (7 km/s)
Burn time	1,680 seconds
Restarts	24

Dimensions

Length	6.9 m (23 ft)
Diameter	2.59 m (8 ft 6 in)
Dry weight	18,144 kg (40,001 lb)

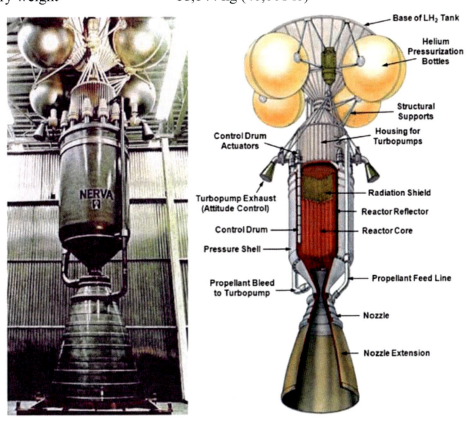

Fig. 2.6 NERVA Diagram. Photo courtesy of NASA Marshall/ M. Houts

The above photo is of a nuclear thermal propulsion (NTP) system from the Rover/NERVA programs and a cutaway schematic with labels.

2.2.1 Nozzle

Nuclear rocket engines are designed to operate at extremely hot temperatures to maximize efficiency. The regenerative cooling system, which flows cold liquid hydrogen through tubes surrounding the nozzle, is an essential element of the design. Unlike chemical rockets, nuclear engines employ a nozzle that narrows sharply before expanding. It was difficult to cool the contraction area. To address this problem, Lewis researchers sought a better understanding of the heat transfer process in the nozzle. They installed experimental copper and steel engines in the J–1 test facility at Plum Brook Station (today, the Neil Armstrong Test Facility). The researchers used the test results from numerous firings of the engine to create a mathematical formula to predict the transfer of heat from the exhaust to the nozzle. They then expanded the investigation by firing the engine with different propellants and injector shapes. The studies at J–1 determined that the injector design needed to be tailored to the shape of the nozzle.

2.2.2 Test Site

Area 25 is the site of the now decommissioned Nuclear Rocket Development Station (NRDS). It was built in support of Project Rover to test prototype nuclear rocket engines. The complex includes three test stands, the Engine Maintenance, Assembly, and Disassembly (E-MAD) facility, the Reactor Maintenance, Assembly, and Disassembly (R-MAD) facility, a control point/technical operations complex, an administrative area and a radioactive material storage area. The R-MAD Facility was built to support the nuclear rocket program and was operational from 1959 through 1970. It was used to assemble reactor engines and to disassemble and study reactor parts and fuel elements after reactor tests. Project Rover was successful, but ultimately canceled.

Fig. 2.7 E-MAD facility in Area 25. Photo courtesy of NNSA Nevada Site Office.

Nuclear reactors for Project Rover were built at LASL Technical Area 18 (TA-18), also known as the Pajarito Site. The reactors were tested at very low power before being shipped to Jackass Flats at the Nevada Test Site. Testing of fuel elements and other materials science was done by the LASL N Division at TA-46 using several ovens and later the Nuclear Furnace.

Work commenced on test facilities at Jackass Flats in mid-1957. All materials and supplies had to be brought in from Las Vegas. Test Cell A consisted of a farm of hydrogen gas bottles and a concrete wall 1 m (3.28 ft) thick to protect the electronic instrumentation from radiation produced by the reactor. The control room was located 3.2 km (2 mi) away. The reactor was test fired with its plume in the air so that radioactive products could be safely dissipated.

The reactor maintenance and disassembly building (R-MAD) was, in most respects, a typical hot cell used by the nuclear industry, with thick concrete walls, lead glass viewing windows, and remote manipulation arms. It was exceptional only for its size: 76 m (250 ft) long, 43 m (140 ft) wide and 19 m (63 ft) high. This allowed the engine to be moved in and out on a railroad car.

The "Jackass and Western Railroad", as it was light-heartedly described, was said to be the world's shortest and slowest railroad. (See Fig. 2.3 above).There were two locomotives, the remotely controlled electric L-1, and the diesel/electric L-2, which was manually controlled but had radiation shielding around the cab. The former was normally used; the latter was provided as a backup. Construction workers were housed in Mercury, Nevada. Later, thirty mobile homes were brought to Jackass Flats to create a village named "Boyerville" after the supervisor, Keith Boyer. Construction work was completed in the fall of 1958. NASA planned to develop a community of 2,700 people, with 800 dwellings and their own shopping complex by 1967.

For a 22 minute video about the Kiwi and NERVA concept, go to:
https://youtu.be/b18HtG0DOCM

IMAGE LINKS

Fig. 2.1 The Simplicity of the Nuclear Thermal Propulsion Rocket
https://www.ans.org/file/9772/Nuclear+Thermal+Propulsion.jpg

Fig. 2.2 Kiwi A Prime
https://www1.grc.nasa.gov/wp-content/uploads/GPN-2002-000141-300x300.jpg

Fig. 2.3 Kiwi B
https://www1.grc.nasa.gov/wp-content/uploads/GRC-1964-C-69681-1024x1056.jpg

Fig. 2.4 Phoebus Nuclear Rocket on the Jackass & Western railroad

https://upload.wikimedia.org/wikipedia/commons/5/5b/Phoebus_nuclear_rock
et_engine.jpg?20190713054959

Fig. 2.5 NERVA
https://wikiimg.tojsiabtv.com/wikipedia/commons/thumb/e/ea/NTS_-_ETS-
1_002.jpg/1280px-NTS_-_ETS-1_002.jpg

Fig. 2.6 NERVA diagram
https://parabolicarc.com/wp-
content/uploads/2021/02/NERVA_image_diagram.jpg

Fig. 2.7 EMAD
https://upload.wikimedia.org/wikipedia/commons/b/bd/NTS_-
_EMAD_Facility_002.jpg

3

Government Efforts

The NASA Nuclear Cryogenic Propulsion Stage (NCPS) project that started in 2011 has evolved over time as a result of a realignment of NASA's current priorities and missions. The effort is now part of a multi-agency alignment discussed below. The traditional NASA Center responsibilities are still in place as are those of the DOD national laboratories.

3.1 DARPA

On January 24, 2023, NASA and the Defense Advanced Research Projects Agency (DARPA) announced a collaboration to demonstrate a nuclear thermal rocket engine in space; an enabling capability for crewed missions to Mars. NASA and DARPA will partner on the Demonstration Rocket for Agile Cislunar Operations (DRACO) program. The non-reimbursable agreement designed to benefit both agencies, outlines roles, responsibilities, and processes aimed at speeding up development efforts.

"NASA will work with our long-term partner, DARPA, to develop and demonstrate advanced nuclear thermal propulsion technology as soon as 2027. With the help of this new technology, astronauts could journey to and from deep space faster than ever; a major capability to prepare for crewed missions to Mars," said NASA Administrator Bill Nelson. "Congratulations to both NASA and DARPA on this exciting investment, as we ignite the future, together."

Using a nuclear thermal rocket allows for faster transit time thereby reducing the risk for the astronauts. Reducing transit time is a key component for human missions to Mars, as longer trips require more supplies and more robust systems. Maturing faster, more efficient transportation technology will help NASA meet its Moon to Mars Objectives.

For a 3:40 minute speech by NASA Administrator Bill Nelson on the DARPA/NASA collaboration, go to: https://www.youtube.com/watch?v=-jmS6pDF3Ho

For a 46:55 minute video on the above collaboration featuring Bill Nelson, Steven D. Howe, Pam Melroy and Stefanie Tompkins, go to: https://www.youtube.com/watch?v=U4D6kx9LRPk

Other benefits to space travel include increased science payload capacity and higher power for instrumentation and communication. In a nuclear thermal rocket engine, a fission reactor is used to generate extremely high temperatures. The engine transfers the heat produced by the reactor to a liquid propellant, which is expanded and exhausted through a nozzle to propel the spacecraft. Nuclear thermal rockets can be three or more times more efficient than conventional chemical propulsion.

"NASA has a long history of collaborating with DARPA on projects that enable our respective missions, such as in-space servicing," said NASA Deputy Administrator Pam Melroy. "Expanding our partnership to nuclear propulsion will help drive forward NASA's goal to send humans to Mars."

Under the agreement, NASA's Space Technology Mission Directorate (STMD) will lead technical development of the nuclear thermal engine to be integrated with DARPA's experimental spacecraft. DARPA is acting as the contracting authority for the development of the entire stage and the engine, which includes the reactor. DARPA will lead the overall program including rocket systems integration and procurement, approvals, scheduling, and security, cover safety and liability, and ensure overall assembly and integration of the engine with the spacecraft. Over the course of the development, NASA and DARPA will collaborate on assembly of the engine before the in-space demonstration as early as 2027.

"DARPA and NASA have a long history of fruitful collaboration in advancing technologies for our respective goals, from the Saturn V rocket that took humans to the Moon for the first time to robotic servicing and refueling of satellites," said Dr. Stefanie Tompkins, director, DARPA. "The space domain is critical to modern commerce, scientific discovery, and national security. The ability to accomplish leap-ahead advances in space technology through the DRACO nuclear thermal rocket program will be essential for more efficiently and quickly transporting material to the Moon and eventually, people to Mars."

The last nuclear thermal rocket engine tests conducted by the United States occurred more than 50 years ago under NASA's Nuclear Engine for Rocket Vehicle Application and Rover projects.

"With this collaboration, we will leverage our expertise gained from many previous space nuclear power and propulsion projects," said Jim Reuter, associate administrator for STMD. "Recent aerospace materials and engineering advancements are enabling a new era for space nuclear technology, and this flight demonstration will be a major achievement toward establishing a space transportation capability for an Earth-Moon economy."

NASA, the Department of Energy (DOE), and industry are also developing advanced space nuclear technologies for multiple initiatives to harness power for space exploration. Through NASA's Fission Surface Power project, DOE awarded three commercial design efforts to develop nuclear power plant concepts that could be used on the surface of the Moon and, later, Mars.

NASA and DOE are working another commercial design effort to advance higher temperature fission fuels and reactor designs as part of a nuclear thermal propulsion engine. These design efforts are still under development to support a longer-range goal for increased engine performance and will not be used for the DRACO engine.

3.1.1 The DRACO Program

The goal of the Demonstration Rocket for Agile Cislunar Operations (DRACO) program is to demonstrate a nuclear thermal rocket (NTR) in orbit. The Program Manager is Dr. Tabitha Dodson. NTRs use a nuclear reactor to heat propellant to extreme temperatures before exhausting the hot propellant through a nozzle to produce thrust. Compared to conventional space propulsion technologies, NTRs offers a high thrust-to-weight ratio around 10,000 times greater than electric propulsion and two-to-five time's greater specific impulse (i.e. propellant efficiency) than in-space chemical propulsion.

- Phase 1 of the DRACO program involved two parallel risk reduction activities.
  Track A conducted a baseline design of an NTR engine. Track B developed an operational system concept to meet operational mission objectives and a demonstration system design that is traceable subsystem.
- Phase 2 will involve a cold flow test of the rocket engine without nuclear fuel.
- Phase 3 of the DRACO program will carry a single design forward to the flight demonstration, which is envisioned to take place by FY27. Phase 3 will involve assembly of the fueled NTR with the stage, environmental testing, and launch into space to conduct experiments on the NTR and its reactor.

Fig. 3.1 DRACO Nuclear Thermal Rocket in flight. Photo courtesy of DARPA

3.2 NASA

3.2.1 Headquarters STMD

NASA's Space Technology Mission Directorate (STMD) at NASA
Headquarters supports the Nuclear Thermal Propulsion efforts as part of the
Game Changing Development (GCD) Program. This Program advances space
technologies that may lead to entirely new approaches for the Agency's future
space missions and provide solutions to significant national needs. GCD
collaborates with research and development teams to progress the most
promising ideas through analytical modeling, ground-based testing and
spaceflight demonstration of payloads and experiments. These teams are held
accountable for ensuring that discoveries move rapidly from the laboratory to

application. GCD's efforts are focused on the mid Technology Readiness Level (TRL) range of (3-5/6), generally taking technologies from initial lab concepts to a complete engineering development prototype.

The Program employs a balanced approach of guided technology development efforts and competitively selected efforts from across academia, industry, NASA, and other government agencies. GCD strives to develop the best ideas and capabilities irrespective of their source.

The Program's investment in innovative space technologies directly supports NASA's mission to "Drive advances in science, technology, and exploration to enhance knowledge, education, innovation, economic vitality, and stewardship of Earth". GCD's focus on transformative space and science technologies will enable science missions and NASA's Artemis Program. Additionally, GCD's technology developments serve as a stimulus to the U.S. economy while providing inspiration and opportunity to our nation's youth.

3.2.2 Marshall Space Flight Center (MSFC)

NASA's Marshall Space Flight Center in Huntsville, Alabama, leads the agency's space nuclear propulsion project in partnership with a DOE team that includes scientists and engineers from Idaho National Laboratory, Los Alamos National Laboratory, and Oak Ridge National Laboratory.

NASA Headquarters' STMD's Technology Demonstration Missions program funds these technology programs. The Science & Technology Office at Marshall strives to apply advanced concepts and capabilities to the research, development and management of a broad spectrum of NASA programs, projects and activities that fall at the very intersection of science and exploration, where every discovery and achievement furthers scientific knowledge and understanding, and supports the agency's ambitious mission to expand humanity's reach across the solar system.

3.2.2.1 NTREES

The Nuclear Thermal Rocket Element Environmental Simulator (NTREES) is an innovative test facility at NASA's Marshall Space Flight Center in Huntsville, AL. It is just one of numerous cutting-edge space propulsion and science research facilities housed in the state-of-the-art Propulsion Research &

Development Laboratory at Marshall, contributing to development of the Space Launch System and a variety of other NASA programs and missions including deep space propulsion missions.

The purpose of the NTREES facility is to perform realistic non-nuclear testing of NTR fuel elements and fuel materials. In an actual reactor, the fuel elements would contain uranium, but no radioactive materials were used during the NTREES tests. Among the fuel options were graphite composite and a "cermet" composite - a blend of ceramics and metals. Both materials were investigated in previous NASA and U.S. Department of Energy research efforts.

Although NTREES cannot mimic the neutron and gamma environment of an operating NTR, it can mimic the thermal hydraulic environment within an NTR fuel element. Once fully operational, NTREES was capable of testing fuel elements and fuel materials in flowing hydrogen at pressures up to 6895 kPa, at temperatures up to and above 3000 K, and at power densities near-prototypic.

NTREES is capable of testing with a variety of propellants, including hydrogen with additives to inhibit corrosion of certain potential NTR fuel forms. The NTREES facility is licensed to test fuels containing depleted uranium. It includes a pyrometer suite to measure fuel temperature profiles and a mass spectrometer to help assess fuel performance and evaluate potential material loss from the fuel element during testing. NTREES is configured to allow continuous testing for any desired length of time, and uses propellant fed from propellant trailers located external to the facility.

The primary NTREES chamber can be reconfigured to allow up to 1 MW to be fed into the fuel element heaters. Future configurations could increase power levels to 5 MW without exceeding the capability of supporting systems.

The NTREES facility also includes an arc heater that has demonstrated the capability to flow hot hydrogen over a material or fuel sample at a hydrogen gas temperature of up to $3160°$ K (5,228°F) for over 30 minutes.

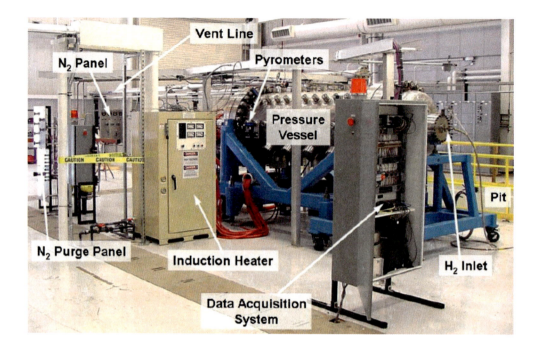

Fig. 3.2 NTREES. Photo courtesy of NASA/Marshall Propulsion Research and Development Laboratory.

The NTREES facility is designed to test fuel elements and materials in hot flowing hydrogen, reaching pressures up to 1,000 pounds per square inch and temperatures of nearly 5,000°F; conditions that simulate space-based nuclear propulsion systems to provide baseline data critical to the research team.

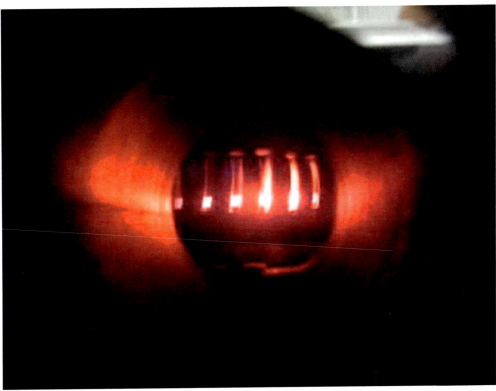

Fig. 3.3 Testing a fuel element. Photo courtesy MSFC/Emmett Given.

This photo was taken in 2012 as a non-nuclear fuel element is heated to more than 3,200 degrees Fahrenheit while hydrogen is funneled through it.

"This is vital testing, helping us reduce risks and costs associated with advanced propulsion technologies and ensuring excellent performance and results as we progress toward further system development and testing," said Mike Houts, Project Manager for nuclear systems at Marshall.

A first-generation nuclear cryogenic propulsion system could propel human explorers to Mars more efficiently than conventional spacecraft, reducing crews' exposure to harmful space radiation and other effects of long-term space missions. It could also transport heavy cargo and science payloads. Further development and use of a first-generation nuclear system could also provide the foundation for developing extremely advanced propulsion technologies and systems in the future, ones that could take human crews even farther into the solar system.

Building on previous, successful research and using the NTREES facility, NASA can safely and thoroughly test simulated nuclear fuel elements of various sizes, providing important test data to support the design of a future Nuclear Cryogenic Propulsion Stage. A nuclear cryogenic upper stage - its liquid-hydrogen propellant chilled to super-cold temperatures for launch - would be designed to be safe during all mission phases and would not be started until the spacecraft had reached a safe orbit and was ready to begin its journey to a distant destination. Prior to startup in a safe orbit, the nuclear system would be cold, with no fission products generated from nuclear operations, and with radiation below significant levels.

"The information we gain using this test facility will permit engineers to design rugged, efficient fuel elements and nuclear propulsion systems," said NASA researcher Bill Emrich, manager of the NTREES facility at Marshall. "It's our hope that it will enable us to develop a reliable, cost-effective nuclear rocket engine in the not-too-distant future."

3.2.3 Glenn Research Center (GRC)

NASA Glenn Research Center is developing a common collaborative full engine simulation tool for the U.S. Government, aerospace industry, and academia called the Numerical Propulsion System Simulation (NPSS). NPSS provides an environment for the analysis and design of propulsion systems for thermodynamic systems. The NPSS focuses on the potential integration of multiple disciplines such as aerodynamics, structures, and heat transfer, along with the concept of numerical zooming between 0-Dimensional to 1-, 2-, and 3- Dimensional component engine codes.

Multiphysics analysis of the reactor, Finite Element Analysis, Computational Fluid Dynamics and Monte Carlo N-Particle Transport (FEA/CFD/MCNP) combines neutronics, fluid, thermal, and structural simulations all coupled together, which captures design subtleties otherwise not seen without coupling the simulations.

Numerical Propulsion System Simulation (NPSS) is an object-oriented, non-linear code originally developed at GRC in 1995 and is used to model full engine systems and integrate model details from our multiphysics analysis. Transient modeling capabilities exist with NPSS in order to model engine startup and shutdown, which is an important use case for NPSS to determine what events occur during the transient engine phases.

Mission Analysis codes developed at GRC are utilized to size vehicle architecture; determine mission timelines, delta-V, mass predictions, and system optimization.

The NTP Materials Data Book is being developed by GRC to better understand high temperature material properties (up to 3000K) in the engine and reactor as these properties are not well understood for fuel elements, moderator elements, and cladding and many data gaps exist at the expected reactor/engine temperatures. To close these data gaps, GRC is leading the necessary thermo-mechanical testing and investigation of fuel coating architectures to reduce the potential for mid-passage corrosion due to hydrogen.

A Laser Test Rig is being developed at GRC in order to better understand and accurately predict the heat transfer from the fuel elements to moderator elements through laser heating of the elements via the high heat flux generated (meant to emulate conditions expected during operation of an NTP engine). The data gained here will help close many of the data gaps at the expected reactor temperatures.

The Heat Transfer Working Group (HTWG) identifies the type of heat transfer data that NTP designer's need develops design standards, and the design of experiments. For instance, one of the key experimental measurements needed is to accurately model heat transfer characteristics from fuel element to moderator element.

For a 7:06 minute video on NASA GRC's work with nuclear fission as applied to the Kilopower Reactor Using Stirling Technology (KRUSTY) go to: https://www.youtube.com/watch?v=0TL7eUh4yuI

3.2.4 Stennis Space Center (SCC)

The John C. Stennis Space Center (SSC) is a NASA rocket testing facility in Hancock County, MI on the banks of the Pearl River at the Mississippi–

Louisiana border. It is NASA's largest rocket engine test facility. There are over 50 local, states, national, international, private, and public companies and agencies using SSC for their rocket testing facilities.

While Stennis is known for testing the largest chemical rockets such as those used on the Saturn, Space Shuttle and Space Launch System, it also has the capability to test smaller rockets, but powerful ones including nuclear thermal propulsion rockets.

3.3 DOE

DOE facilities supporting NASA's space nuclear propulsion project include the Idaho National Laboratory, Oak Ridge National Laboratory, and Los Alamos National Laboratory.

Based on substantial experimental databases and anticipated performance potential, high-temperature graphite fuels and tungsten CERMET fuels have been shown to be the two near-term candidate fuels that are capable of meeting the operational requirements for NTP systems. Fuel expertise and fabrication facilities exist at the Idaho National Laboratory (INL), Oak Ridge National Laboratory (ORNL), and Los Alamos National Laboratory (LANL). While, new fabrication equipment would have to be procured and installed for both the graphite and CERMET fuels, substantial fuel fabrication and evaluation capabilities exist within the Department of Energy (DOE) complex to support the fuel development. The following describes what the Laboratories are doing to support nuclear thermal propulsion efforts.

3.3.1 Idaho National Laboratory (INL)/Battelle Energy Alliance

The contracts, to be awarded through the DOE's Idaho National Laboratory (INL), are each valued at approximately $5 million. They fund the development of various design strategies for the specified performance requirements that could aid in deep space exploration. "By working together, across government and with industry, the United States is advancing space nuclear propulsion," said Jim Reuter, Associate Administrator for NASA's Space Technology Mission Directorate. "These design contracts are an important step towards tangible reactor hardware that could one day propel new missions and exciting discoveries."

Battelle Energy Alliance, the managing and operating contractor for INL, led the request for proposals, evaluation, and procurement sponsored by NASA using fiscal year 2021 appropriations. INL will award 12-month contracts to the following companies to each produce a conceptual reactor design that could support future mission needs:

- BWX Technologies, Inc. of Lynchburg, Virginia – The company will partner with Lockheed Martin.
- General Atomics Electromagnetic Systems of San Diego – The company will partner with X-energy LLC and Aerojet Rocketdyne.
- Ultra Safe Nuclear Technologies of Seattle – The company will partner with Ultra Safe Nuclear Corporation, Blue Origin, General Electric Hitachi Nuclear Energy, General Electric Research, Framatome, and Materion.

See Chapter 4 for more details on each company.

"INL is excited to enable the development of nuclear propulsion technology for potential use by NASA in future space exploration," said Dr. Stephen Johnson, national technical director for space nuclear power and director of the Space Nuclear Power and Isotope Technologies Division at INL. "Our national laboratories, working in partnership with industry, bring unparalleled expertise and capabilities to assist NASA in solving highly complex challenges that come with nuclear power and propulsion."

At the end of the contracts' performance periods, INL will conduct design reviews of the reactor concepts and provide recommendations to NASA. NASA will utilize the information to establish the basis for future technology design and development efforts. NASA is also maturing a fission surface power system for use on the Moon and Mars. NASA intends to partner with the DOE and INL to release a request for proposals that asks industry for preliminary designs of a 10-kilowatt class system that NASA could demonstrate on the lunar surface. Maturing fission surface power can also help inform nuclear electric propulsion systems, another candidate propulsion technology for distant destinations.

3.3.1.1 TREAT

The Transient Reactor Test Facility (TREAT) is a national asset that provides unique test results in an essential nuclear research field. It will foster the development of new ways to provide baseload and load following electrical power. Transient testing is an essential component of the United States and international efforts to develop robust, safer nuclear fuels, and to bring innovative reactor technologies to the market.

Transient testing involves the application of controlled, short-term bursts of intense neutron flux directed toward a test specimen in order to study fuel and material performance under off-normal operational conditions and hypothetical accident scenarios. After the transient test, the fuel or material is analyzed at a post-irradiation examination (PIE) facility. The results of these examinations are then evaluated and used in advancing fuel or material design and qualification.

TREAT is a highly capable test reactor. Detailed real-time monitoring of the specimens during a test is possible via the hodoscope, a system that detects fast neutron signatures from experiments, and other experiment and core instrumentation. This instrumentation, coupled with PIE, allows scientists to determine the appropriate safety limits for the fuels and materials in nuclear power reactors. TREAT's simple, self-limiting, air-cooled design can safely accommodate multipin test assemblies, enabling the study of fuel melting, metal-liquid reactions, overheated fuel and coolant reactions, and transient behavior of fuels for high temperature system applications. It studies the potential for propagation of failure to adjacent fuel pins under conditions ranging from mild upsets to severe accidents.

The TREAT facility operated from 1959 through 1994, when it was placed in standby mode. A resurgence of interest in developing innovative nuclear technologies has driven demand for transient testing. TREAT was restarted in 2018 and is currently supporting experiment programs.

The reactor was originally constructed to test fast-reactor fuels, but its flexible design has also enabled its use for testing of light-water-reactor fuels as well as other exotic special-purpose fuels, such as space reactors. TREAT has an open-core design that allows for ease of experiment instrumentation and real-time imaging of fuel motion during irradiation, which also makes TREAT an ideal platform for understanding the irradiation response of materials and fuels on a fundamental level.

Basic Capabilities:

- High-intensity (20 GW), short-duration (<100 ms) neutron pulses for severe accident testing,
- Shaped transients at intermediate powers and times (flexible power shapes with up to several minutes duration), 120 kW steady state operation,
- Testing capability for static capsules, sodium loops, water loops and hydrogen loops and,
- Neutron radiography facility.

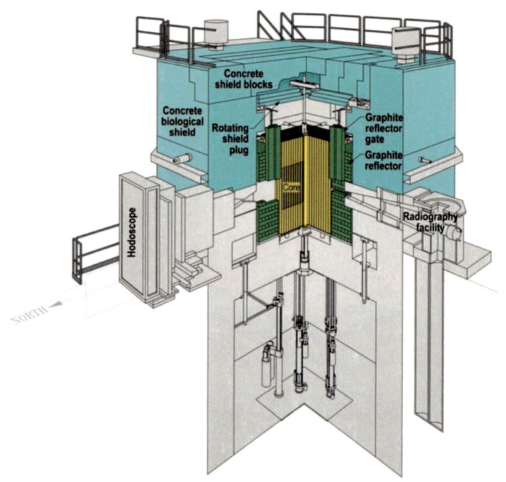

Fig. 3.4 The TREAT Test Facility Reopened in 2018. Photo courtesy of INL.

Fig. 3.5 Annotated view of the TREAT. Photo courtesy of INL>

For a 1:06 minute video of the TREAT, go to: https://youtu.be/x-ARTzPj6os

3.3.2 Oak Ridge National Laboratory

Oak Ridge National Laboratory (ORNL) is participating in the nuclear thermal propulsion (NTP) research and development effort supported by NASA. This effort involves collaboration between multiple research groups, representing various government agencies and industry partners. ORNL has developed a Modelica-based modeling package for dynamic system modeling of nuclear reactors called the Transient Simulation Framework of Reconfigurable Models (TRANSFORM). While this software has been successfully demonstrated in simulations of traditional pressurized-water reactors, boiling-water reactors, liquid-metal reactors and molten-salt reactors, it has also been adapted for non-traditional use in modeling hybrid energy systems, and tritium transport. This versatility is being applied to the current NTP project, where specific modules within TRANSFORM are being developed for and applied to transient modeling of the NASA NTP design. This effort presents the current state of the ORNL-NTP model, the utility of TRANSFORM methodologies in NTP transient simulations, the ability to develop NTP-specific modules, and future work for the model.

3.3.3 Los Alamos National Laboratory (LANL)

Los Alamos National Laboratory (LANL), a multidisciplinary research institution engaged in strategic science on behalf of national security, is managed by Triad, a public service oriented, national security science organization equally owned by its three founding members: Battelle Memorial Institute (Battelle), the Texas A&M University System (TAMUS), and the Regents of the University of California (UC) for the Department of Energy's National Nuclear Security Administration.

Los Alamos enhances national security by ensuring the safety and reliability of the U.S. nuclear stockpile, developing technologies to reduce threats from weapons of mass destruction, and solving problems related to energy, environment, infrastructure, health, and global security concerns.

LANL and its contractors supported NASA on the Rover and NERVA projects providing the basic reactor design, fuel materials development, and reactor testing capability. NERVA meanwhile was focused on engine development by the industrial team of Aerojet and Westinghouse, building on and extending the Los Alamos efforts for flight system development. The challenge of this part of the program was designing nuclear engines that could survive the shock and vibration of a space launch. LANL is currently teamed with NASA/DOD on new efforts to study Nuclear Thermal Propulsion for deep space missions as well as to produce nuclear power for lunar and Mars activities.

LANL has signed an agreement to license its "Kilopower" space reactor technology to Space Nuclear Power Corporation (SpaceNukes) also based in Los Alamos, NM. See Chapter 4.2)

"We developed this technology at the Laboratory in partnership with NASA and the National Nuclear Security Administration," said Patrick McClure, who served as project lead for Kilopower at Los Alamos and is now a partner in SpaceNukes. "By creating our own company, we're hoping to be able to reach potential new sponsors who will want to take this technology to the next level and put it into space."

Kilopower is a small, lightweight fission power system capable of providing various ranges of power depending on the need. For example, SpaceNukes offers low-kilowatt reactors to power deep space missions, middle-range reactors in the tens of kilowatts to power a lunar or Martian habitat, and much larger reactors in the hundreds of kilowatts that could make enough propellant for a rocket to return to Earth after a stay on Mars.

IMAGE LINKS

Fig. 3.1 DRACO in flight
https://www.nasa.gov/sites/default/files/styles/full_width/public/thumbnails/image/draco-4-darpa-nasa-ussf-caption.jpg?itok=6b5V9gGv

Fig. 3.2 Annotated NTREES Test Facility
https://www.researchgate.net/publication/255240279/figure/fig3/AS:29797143
5491332@1448053334932/Nuclear-Thermal-Rocket-Element-Environmental-Simulator-NTREES.png

Fig. 3.3 NTREES Fuel Element Test
https://encrypted-tbn1.gstatic.com/images?q=tbn:ANd9GcTuQSTTvObiv2u-pKiYvHnxB0rbzv01L_w2CeW1q-21tc5c2VxI

Fig. 3.4 the TREAT facility
https://mfc-inlgov.imgix.net/SiteCollectionImages/SitePages/Transient%20Reactor%20Test%20Facility/TREAT.png?auto=compress,enhance&fit=scale&w=450

Fig. 3.5 Annotated diagram of the TREAT facility
https://www.osti.gov/biblio/1656842/image/008/809/0088097/2/t0485728.png

4

Nuclear Thermal Propulsion Contractors

Background

On July 13, 2021, NASA Announced the Nuclear Thermal Propulsion Reactor Concept Awards. Working with the Department of Energy (DOE), NASA is leading an effort to advance space nuclear technologies. The government team has selected three reactor design concept proposals for a nuclear thermal propulsion system. The reactor is a critical component of a nuclear thermal engine, which would utilize high-assay low-enriched uranium fuel (HALEU).

The 12 month contracts were awarded through the DOE's Idaho National Laboratory (INL); each valued at approximately $5 million. They funded the development of various design strategies for the specified performance requirements that could aid in deep space exploration. "INL is excited to enable the development of nuclear propulsion technology for potential use by NASA in future space exploration," said Dr. Stephen Johnson, national technical director for space nuclear power and director of the Space Nuclear Power and Isotope Technologies Division at INL. "Our national laboratories, working in partnership with industry, bring unparalleled expertise and capabilities to assist NASA in solving highly complex challenges that come with nuclear power and propulsion."

Battelle Energy Alliance (BEA), the managing and operating contractor for INL, led the request for proposals, evaluation, and procurement sponsored by NASA using fiscal year 2021 appropriations. INL awarded the contracts to the following three companies and their teams to each produce a conceptual reactor design that could support future mission needs.

At the end of the contracts' performance periods, INL conducted design reviews of the reactor concepts and provide recommendations to NASA. NASA utilized the information to establish the basis for future technology design and development efforts. This led to the NASA/DOD/DARPA collaboration discussed in Chapter 3.

4.1 THE BWX TECHNOLOGIES TEAM

BWX Technologies, Inc., headquartered in Lynchburg, VA is a supplier of nuclear components and fuel to the U.S. On July 1, 2015, BWX Technologies Inc. began trading separately from its former subsidiary Babcock & Wilcox Enterprises Inc. after a spinoff.

While its founders go back to the 1850's, their involvement in Nuclear Thermal Propulsion goes back to a contract in1987 and 1993 patent for solar bi-modal space power and propulsion system. In 1999, they were awarded a contract to develop fuel cells and hydrogen reformers for the U.S. Navy. In 2017, BWXT was awarded $18.8 million contract from NASA to initiate conceptual designs for a nuclear thermal propulsion reactor. While the company is involved with many nuclear related projects, this section will concentrate on the more recent NASA/DOD/INL/DARPA contract efforts.

On April 1, 2021 BWX Technologies, Inc. announced that it is continuing its groundbreaking Nuclear Thermal Propulsion (NTP) design, manufacturing development, and test support work for NASA. NTP is one of the technologies that are capable of propelling a spacecraft to Mars; this contract continues BWXT's work that began in 2017.

Under the terms of a $9.4 million, one-year contract awarded to its BWXT Advanced Technologies subsidiary, BWXT will focus primarily on nuclear fuel design and engineering activities. Specifically, BWXT will produce fuel kernels, coat the fuel kernels, design materials and manufacturing processes for fuel assemblies, and further develop conceptual reactor designs, among other activities. The work will be conducted primarily at BWXT's Advanced Technology Laboratory, Specialty Fuel Facility, and Lynchburg Technology Center. It will involve more than 50 employees.

"BWXT has a decades-long history of supporting NASA, and we are very proud to continue our efforts for the country," said Ken Camplin, BWXT Nuclear Services Group president. "Our designers and engineers have been working with teams at NASA, the DOE National Laboratories and academia to help the United States accomplish one of humankind's ultimate goals: to send astronauts to another planet and return them safely. This is exciting work for us, and it demonstrates the incredible diversity of talent and expertise that our company has developed."

BWXT has been making significant progress on NASA's NTP initiative, which has progressed from the Space Technology Mission Directorate's Game Changing Development program to its Technology Demonstration Mission program. BWXT's progress to date includes evaluating various fission fuel and reactor options, developing a conceptual reactor design, tailoring the fuel design to use High Assay Low Enriched Uranium (HALEU), and delivering specialty fuel particles for testing.

Fig. 4.1 BWXT Fuel Element in a Glove Box. Photo courtesy of BWXT.

In 2020, as part of NASA's in-space demonstration mission, BWXT delivered a study exploring several reactor configurations and fuel forms capable of delivering space nuclear propulsion. Two of the designs focused on power levels suitable for space demonstration in the near term. A third design was developed that leverages more advanced technology and higher power levels that could be ready in time for a Mars mission.

Rocket engines based on NTP technology are designed to propel a spacecraft from Earth orbit to Mars and back. Nuclear Thermal Propulsion for spaceflight has a number of advantages over chemical-based designs. In particular, NTP provides a low mass capability that allows astronauts to travel through space faster, thereby reducing supply needs and lowering their exposures to cosmic radiation.

On December 13, 2021, BWXT reached a critical milestone in the nation's pursuit of space nuclear propulsion by delivering coated reactor fuels to NASA in support of its space nuclear propulsion project. BWXT has been able to leverage its decades of specialty and coated fuel manufacturing experience as well as its existing licensed production facilities to be the first private company to deliver relevant coated fuels that will be used in NASA testing scheduled in the future.

Under the terms of a previously announced contract awarded to BWXT by Idaho National Laboratory, the company will continue to produce fuel kernels, coated fuel kernels, and design materials and manufacturing processes for fuel assemblies. BWXT is developing two fuel forms in support of a reactor ground demonstration by the late 2020s. BWXT is also designing a third, more advanced and more energy-dense fuel form that could be evaluated in the future. BWXT produces a variety of fuels that enable diverse mission concepts from high-temperature coated fuels for space exploration to Tri-structural Isotropic (TRISO) fuels for terrestrial use in micro reactors.

In addition to TRISO, BWXT also produces specialty coated fuels for NASA in support of its space nuclear propulsion project within the agency's Space Technology Mission Directorate. See also Chapter 6.5 for more on TRISO fuels.

4.1.1 Lockheed Martin

Lockheed Martin's space nuclear systems work includes three current contracts; a partnership with BWXT Technologies on both nuclear thermal reactor and fission surface power concepts for NASA and DOE and a contract with DARPA to develop a spacecraft concept design with NTP capability.

While nuclear systems for deep space are an emerging field, Lockheed Martin has a long history and expertise in nuclear controls, having supported instrumentation and controls for both terrestrial power plants and Naval nuclear reactors. Lockheed Martin will apply its expertise in avionics, mission control and integration to this NTP effort.

"Our discriminator also comes from our deep space exploration heritage, which requires the ability to do high-technology, first-of-a-kind missions," said said Lisa May, Principal Engineer and Next Gen Strategy Lead, Lockheed Martin. "We've also invested heavily in cryogenic hydrogen storage and transfer, as well as the overall nuclear reactor controls."

The company also built the radioisotope thermoelectric generators for NASA planetary missions such as Viking, Pioneer, Voyager, Apollo, Cassini and New Horizons.

4.2 THE GENERAL ATOMICS ELECTROMAGNETIC SYSTEMS TEAM

General Atomics Electromagnetic Systems (GA-EMS) supported the DRACO program through a contract awarded by DARPA in April 2021 which concluded in 2022. This effort is now completed. General Atomics completed a baseline design of a nuclear thermal propulsion reactor and engine and tested the components of the reactor as part of the initial phase of the DARPA Demonstration Rocket for Agile Cislunar Operations program. On November 8, 2022, the company assessed the high-temperature fuel elements of the NTP engine at NASA's Nuclear Thermal Rocket Element Environment Simulator (NTREES) as part of the DRACO program's Track A.

"We have leveraged our expertise in nuclear and space system technologies to design an NTP system and test the vital components of that system to confirm they will withstand the relevant design conditions," said Scott Forney, president of GA-EMS. "Unlike electric and chemical propulsion technologies in use today, NTP propulsive capabilities can achieve two to three times the propellant mass efficiency, which is critically important for cislunar missions," Forney added.

The company partnered with X-energy LLC and Aerojet Rocketdyne.

4.2.1 X-energy LLC

X-energy is an American private nuclear reactor and fuel design engineering company. The company was founded in 2009 by Kam Ghaffarian. Since its founding it has received various government grants and contracts, notably through the Department of Energy's (DOE) Advanced Reactor Concept Cooperative Agreement in 2016 and its Advanced Reactor Demonstration Program (ARDP) in 2020. In 2019, X-energy received funding from the DOE to develop small military reactors for use at forward bases.

It is developing a Generation IV high-temperature gas-cooled pebble-bed nuclear reactor design. In October 2020, the company was chosen by the DOE as a recipient of a matching grant totaling between $400 million and $4 billion over the next 5 to 7 years for the cost of building a demonstration reactor of their Xe-100, helium-cooled pebble-bed reactor design. DOE also awarded the same grant to TerraPower.

The Xe-100 is a pebble bed high-temperature gas-cooled nuclear reactor design that is planned to be smaller, simpler and safer when compared to conventional nuclear designs. Pebble bed high temperature gas-cooled reactors were first proposed in 1944. Each reactor is planned to generate 200 MWt and approximately 76 MWe. The fuel for the Xe-100 is a spherical fuel element, or pebble that utilizes the tri-structural isotropic (TRISO) particle nuclear fuel design, with Uranium enriched to 20%, to allow for longer periods between refueling. X-energy claims that TRISO fuel will make nuclear meltdowns virtually impossible.

In the summer of 2020, X-energy submitted its concepts for a nuclear thermal propulsion reactor capable of achieving a specific impulse of 900 seconds. X-energy's design for a nuclear thermal propulsion system would be capable of more than twice the specific impulse of the Saturn V.

A big question for NASA is what type of nuclear fuel design to employ in these planetary exploration reactors. Concerns about proliferation risks have dampened enthusiasm for using high enriched uranium, but the low-enriched uranium used by existing terrestrial nuclear reactors lacks the energy density to meet the needs of a high-temperature propulsion reactor system. High-assay low enriched uranium (HALEU) fuel, which occupies a middle ground between low- and high-enriched uranium (up to 20% enriched), is a strong contender for nuclear thermal rockets.

X-energy is one of the only companies in the U.S. capable of producing ceramic-coated fuel forms using HALEU, which is at the core of our TRISO fuel. Each TRISO fuel kernel consists of a 0.5 micron pellet of uranium oxycarbide; the size of a poppy seed, wrapped in three alternating layers of graphite and silicon carbide. Thousands of these particles are embedded in a graphite fuel form: either pebbles or prismatic compacts. In X-energy's terrestrial reactor, the Xe-100, more than 60,000 of these pebbles (roughly the size of a cue ball) will be cycled through the reactor core over the course of a year.

The most critical element in the design of advanced nuclear reactors is a robust fuel that can withstand very high temperatures without melting. X-energy reactors use tri-structural isotropic (TRISO) particle fuel, developed and improved over 60 years. They manufacture their own proprietary version, (TRISO-X), to ensure supply and quality control.

4.2.2 Aerojet Rocketdyne

NASA is looking into reducing risk and increasing the feasibility of nuclear propulsion for human-rated missions to Mars. In cooperation with NASA and other industry partners, Aerojet Rocketdyne has been leading the research effort on engine and mission architecture planning for this effort.

Nuclear thermal propulsion (NTP) has tremendous synergy with existing liquid rocket technologies; an area in which Aerojet Rocketdyne is a proven leader. This includes the development/production of turbomachinery and nozzles, cryogenic hydrogen fuel handling and providing heat management solutions.

Nuclear electric propulsion (NEP) relies on several key technologies, including high-power electric thrusters, power processing, power conversion, and power management and distribution, which are all areas in which Aerojet Rocketdyne is an established industry leader.

Aerojet Rocketdyne is working with manufacturers to leverage new low-enriched Uranium technology that is safer to handle and reduces expensive regulatory and security challenges; making both NTP and NEP attractive alternatives to comparable chemical and solar electric propulsion options for crewed deep space missions.

4.3 THE USNC TECHNOLOGIES TEAM

Ultra Safe Nuclear Corporation (USNC)-Technologies, a U.S. corporation headquartered in Seattle, WA is a global leader and vertical integrator of nuclear technologies and services, on Earth and in Space. Major initiatives include the Micro Modular Reactor (MMR®), Fully Ceramic Micro-encapsulated (FCM®) nuclear fuel, and nuclear power and propulsion technologies for space exploration.

Idaho National Laboratory has selected USNC-Tech and its partners to develop a nuclear thermal propulsion (NTP) reactor concept design for space exploration: the Power-Adjusted Demonstration Mars Engine (PADME) NTP engine. This effort, one of three selected by the government team, is a step toward the manufacture and demonstration of safe, affordable, reliable, high-performance NTP engines for crewed deep space travel. In the future, the designs could inform a full-scale NTP engine prototype. The funding for this procurement was provided by NASA.

"With PADME, we are making design choices that minimize technical risk and development time," stated Dr. Michael Eades, Director of Engineering at USNC-Tech. "By the end of the decade, NTP will give humanity a platform for doing incredible things beyond low earth orbit, with far greater mobility and flexibility than we've ever had before."

"USNC-Tech started envisioning HALEU-fueled, NTP-powered crewed missions to Mars over 9 years ago, and this award is a tangible step to making that vision a reality," stated Dr. Paolo Venneri, CEO of USNC-Tech and Executive Vice President of USNC's Advanced Technologies Division. "We've assembled an exceptional team and are working with world-class partners."

Based in Seattle, WA, Ultra Safe Nuclear (USNC) will provide nuclear hardware and services for reliable energy anywhere; on Earth and in Space. From milliWatts to megawatts, Ultra Safe Nuclear Corporation Technologies

(USNC-Tech) is developing nuclear power and propulsion technologies to support the sustainable exploration and development of space. Together, the family of USNC divisions offers full-stack capabilities from specialized reactor design services, to advanced materials and fuel development, to reactor manufacturing and system deployment.

Supported by Blue Origin, GE Hitachi Nuclear Energy, GE Research, Framatome, and Materion, the USNC-Tech team is focused on a design approach that supports manufacturability, ease of integration and deployment, mission expansion and ultimately commercial viability.

4.3.1 Blue Origin

Ultra Safe Nuclear Technologies is partnering with Blue Origin on a $5 million contract from NASA and the Department of Energy to develop a reactor design for space-based nuclear thermal propulsion systems.

Blue Origin had a separate $2.5 million DARPA contract to work on the first phase of a program aimed at demonstrating a nuclear thermal propulsion system.

4.3.2 GE Hitachi Nuclear Energy

GE Hitachi Nuclear Energy is a provider of advanced reactors and nuclear services. It is headquartered in Wilmington, North Carolina. Established in June 2007, GEH is a nuclear alliance created by General Electric and Hitachi. In Japan, the alliance is Hitachi-GE Nuclear Energy.

4.3.3 Framatome

Framatome is a French nuclear reactor business. It is owned by Électricité de France, Mitsubishi Heavy Industries, and Assystem. The company first formed in 1958 to license Westinghouse's pressurized water reactor designs for use in France.

Framatome's North American headquarters are in Lynchburg, VA; also the home of BWX Technologies. It has a strong presence in the U.S. nuclear energy market, helping power 36 million American homes. As a reliable partner with a long history of proven performance, they focus on servicing and fueling the U.S. operating nuclear fleet, supporting secondary license renewal as well new nuclear reactor development, and advancing the future of nuclear energy here and abroad.

Framatome and its predecessor organizations have been serving the nuclear energy industry in the United States since the 1950s. Over time, we have serviced every nuclear energy facility in the United States. We are honored to be recognized year after year for our innovative solutions, thanks to the commitment and expertise of our employees.

4.3.4 Materion

Materion Global Headquarters are located Mayfield Heights, OH. Its founder, Brush Beryllium started in 1931 and has a long relationship with the government, the AEC and the DOD. It is known for its supplier of unique materials. The scientists working on the James Webb Space Telescope decided beryllium was perfectly suited to this mission because of its strong, yet light weight, properties and its ability to be polished; important characteristics when making high tech mirrors.

Materion's nuclear grade beryllium is the reference grade for the International Thermonuclear Experimental Reactor (ITER), a cutting-edge fusion energy program. S-65 beryllium metal was chosen for the walls in ITER and for breeder pebbles because of its ability to resist cracking under high heat flux thermal cycling. They were also used for nuclear reflector tiles for the walls in the Joint European Torus (JET), the world's largest plasma physics experiment focused on nuclear fusion energy production. Other applications include neutron reflectors and moderators, neutron filter assemblies, nuclear test reactors and medical isotope reactors.

Nuclear thermal propulsion development is also supported by the Massachusetts Institute of Technology, University of Alabama Huntsville, the Aerospace Corporation, Analytical Mechanics Associates, and Geocent. Each supporting entity brings its own unique expertise and capabilities to contribute to the goal of realizing a high-performance fission-based propulsion system to enable extended human exploration of the solar system.

IMAGE LINKS

Fig. 4.1 Coated uranium kernels viewed thru a glove box
https://www.ans.org/file/5395/BWXT%20coated%20uranium%20fuel%20kernels.jpg

Fig. 4.2 Fission Surface Power BWXT
https://www.lockheedmartin.com/content/dam/lockheed-martin/space/photo/ntp/ntp.jpg.pc-adaptive.1280.medium.jpg

5

Nuclear Fission Surface Power Contractors

5.1 HISTORY OF PAST EFFORTS

To have an understanding as to what led up to these recent contracts involving fission surface power is to better understand the previous efforts. From 2015 to 2018, the "Kilopower" experimental project aimed at producing new nuclear reactors for space travel including surface power for the Moon and Mars. The project started in October, 2015 and was led by NASA and the DOE's National Nuclear Security Administration (NNSA). As of 2017, the Kilopower reactors were intended to come in four sizes, able to produce from one to ten kilowatts of electrical power (1-10 kWe) continuously for twelve to fifteen years. The fission reactor used uranium-235 to generate heat that is carried to Stirling converters with passive sodium heat pipes. In 2018, positive test results from the Kilopower Reactor Using Stirling Technology (KRUSTY) demonstration reactor were achieved.

Potential applications include nuclear electric propulsion and a steady electricity supply for crewed or robotic space missions that require large amounts of power, especially where sunlight is limited or not available. NASA has also studied the Kilopower reactor as the power supply for crewed Mars missions.

Fig. 5.1 Kilopower on Mars. Conceptual artistic rendering courtesy of NASA

During those missions, the reactor would be responsible for powering the machinery necessary to separate and cryogenically store oxygen from the Martian atmosphere for ascent vehicle propellants. Once humans arrive, the reactor would power their life-support systems and other requirements. NASA studies have shown that a 40 kWe reactor would be sufficient to support a crew of between 4 and 6 astronauts.

The reactor is fueled by an alloy of 93% uranium-235 and 7% molybdenum. The core of the reactor is a solid cast alloy structure surrounded by a beryllium oxide reflector, which prevents neutrons from escaping the reactor core and allows the chain reaction to continue. See Chapter 6.8. The reflector also reduces the emissions of gamma radiation that could impair on-board electronics. A uranium core has the benefit of avoiding uncertainty in the supply of other radioisotopes, such as plutonium-238 whose supply is restricted. Plutonium is currently used on Mars rovers' radio isotope thermal generators (RTGs).

The prototype KRUSTY 1 kWe Kilopower reactor weighs 134 kg (295 lbs) and contains 28 kg (62 lbs) of U235. The space-rated 10 kWe Kilopower for

Mars is expected to weigh 1500 kg (3307 lbs) in total (with a 226 kg (498 lbs) core) and contain 43.7 kg (96 lbs) of U235.

Nuclear reaction control is provided by a single rod of boron carbide, which is a neutron absorber. The reactor is intended to be launched cold, preventing the formation of highly radioactive fission products. Once the reactor reaches its destination, the neutron absorbing boron rod is removed to allow the nuclear chain reaction to start. Once the reaction is initiated, decay of a series of fission products cannot be stopped completely. However, the depth of control rod insertion provides a mechanism to adjust the rate at which uranium fissions, allowing the heat output to match the load.

Passive heat pipes filled with liquid sodium transfer the reactor core heat to one or more free-piston Stirling engines, which produce reciprocating motion to drive a linear electric generator. The melting point of sodium is 98 °C (208 °F), which means that liquid sodium can flow freely at high temperatures between about 400 and 700 °C (750 and 1,300 °F). Nuclear fission cores typically operate at about 600 °C (1,100 °F).

The reactor is designed to be intrinsically safe in a wide range of environments and scenarios. Several feedback mechanisms are employed to mitigate a nuclear meltdown. The primary method is passive cooling, which requires no mechanical mechanisms to circulate coolant. The reactor design is self-regulating through design geometry that creates a negative temperature reactivity coefficient. In effect this means that as the power demand increases the temperature of the reactor drops. This causes it to shrink, preventing neutrons from leaking out which in turn causes reactivity to increase and power output to increase to meet the demand. This also works in reverse for times of lower power demand.

The development of Kilopower began with an experiment called DUFF or Demonstration Using Flattop Fissions, (often called the Father of Kilopower) which was tested in September, 2012 using the existing Flattop assembly as a nuclear heat source. When DUFF was tested at the Device Assembly Facility at the Nevada Test Site, it became the first Stirling engine powered by fission energy and the first use of a heat pipe to transport heat from a reactor to a power conversion system. According to David Poston, the leader of the Compact Fission Reactor Design Team, and Patrick McClure, the manager for small nuclear reactor projects at Los Alamos National Laboratory, the DUFF experiment showed that "for low-power reactor systems, nuclear testing can be accomplished with reasonable cost and schedule within the existing infrastructure and regulatory environment".

The prototype Kilopower uses a solid, cast uranium-235 reactor core, about the size of a paper towel roll. Reactor heat is transferred via passive sodium heat pipes, with the heat being converted to electricity by Stirling engines. Testing to gain technology readiness level (TRL) 5 started in November 2017 and continued into 2018. The testing of KRUSTY represents the first time the U.S. has conducted ground tests on any space reactor since the SNAP-10A experimental reactor was tested and eventually flown in 1965.

The KRUSTY reactor was run at full power on March 20, 2018 during a 28-hour test using a 28 kg (62 lbs) U235 reactor core. A temperature of 850 °C (1,560 °F) was achieved, producing about 5.5 kW of fission power. The test evaluated failure scenarios including shutting down the Stirling engines, adjusting the control rod, thermal cycling, and disabling the heat-removal system.

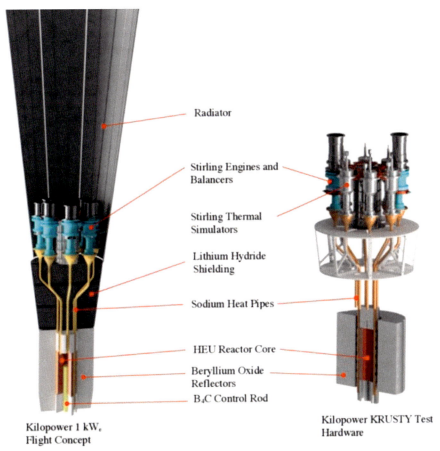

Fig. 5.2 Kilopower System Annotated. Illustration courtesy of NASA.

The Kilopower team conducted the experiment in four phases. The first two phases, conducted without power, confirmed that each component of the system behaved as expected. During the third phase, the team increased power to heat the core incrementally before moving on to the final phase. The experiment culminated with a 28-hour, full-power test that simulated a mission, including reactor startup, ramp to full power, steady operation and shutdown.

According to David Poston, the chief reactor designer at NNSA's Los Alamos National Laboratory, the purpose of the recent experiment in Nevada was two-fold: to demonstrate that the system can create electricity with fission power, and to show the system is stable and safe no matter what environment it encounters. "We threw everything we could at this reactor, in terms of nominal and off-normal operating scenarios and KRUSTY passed with flying colors," said Poston.

The Kilopower project is developing mission concepts and performing additional risk reduction activities to prepare for a possible future flight demonstration. The project will remain a part of the STMD's Game Changing Development program with the goal of transitioning to the Technology Demonstration Mission program in the future. NASA's fission surface power project is managed by NASA's Glenn Research Center. The technology development and demonstration are funded by the Space Technology Mission Directorate's Technology Demonstration Missions program, which is hosted at Marshall Space Flight Center.

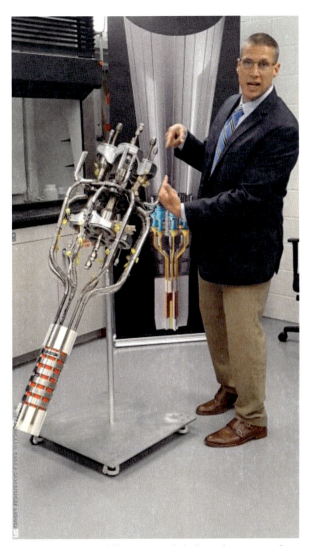

Fig. 5.3 Marc Gibson explaining the operation of the KRUSTY prototype design. Photo courtesy of Michael Cole / Spaceflight Insider

Marc Gibson, NASA Glenn's Kilopower Lead Engineer, explains the operation of the KRUSTY Kilopower prototype design to members of the media at NASA Glenn's Sterling Research Laboratory. The unit can generate up to 10 kilowatts of power continuously over decades with minimal maintenance.

For a 3:16 minute video of Kilopower, go to: https://youtu.be/DcdfMcjUy_U
For a 7:06 minute video narrated by Marc Gibson, go to:
https://www.youtube.com/watch?v=0TL7eUh4yuI
For a 52 page Power Point on the Kilopower Project, go to:
https://ntrs.nasa.gov/api/citations/20180005553/downloads/20180005553.pdf

5.2 CURRENT EFFORTS

On June 21, 2022 NASA Announced the Artemis Concept Awards for Nuclear
Power on the Moon. NASA and the DOE are working together to advance
space nuclear technologies. The agencies have selected three design concept
proposals for a fission surface power system design that could be ready to
launch by the end of the decade for a demonstration on the Moon. This
technology would benefit future exploration under the Artemis umbrella.

The contracts, to be awarded through the DOE's Idaho National Laboratory,
are each valued at approximately $5 million. The contracts fund the
development of initial design concepts for a 40-kilowatt class fission power
system planned to last at least 10 years in the lunar environment.

Relatively small and lightweight compared to other power systems, fission
systems are reliable and could enable continuous power regardless of location,
available sunlight, and other natural environmental conditions. A
demonstration of such systems on the Moon would pave the way for long-
duration missions on the Moon and Mars.

"New technology drives our exploration of the Moon, Mars, and beyond,"
said Jim Reuter, Associate Administrator for NASA's Space Technology
Mission Directorate. "Developing these early designs will help us lay the
groundwork for powering our long-term human presence on other worlds."

Battelle Energy Alliance, the managing and operating contractor for Idaho
National Laboratory, led the Request for Proposal development, evaluation,
and procurement sponsored by NASA. Idaho National Laboratory awarded 12-
month contracts to the following companies to each develop preliminary
designs.

"At the end of the period, NASA will be looking at going into phase two, which is a lot more funding and developing hardware, as well as proving out the designs. So right now it's purely a paper design and a study by all three teams," said Dr. Stephen G. Johnson, Director of the Space Nuclear Power and Isotope Technologies Division at the INL Materials and Fuels Complex. After the 12-month study, INL will determine the best contract and then move forward with building the reactor for launch.

"The Fission Surface Power project is a very achievable first step toward the U.S. establishing nuclear power on the Moon," said Idaho National Laboratory Director John Wagner. "I look forward to seeing what each of these teams will accomplish."

The Phase 1 awards will provide NASA critical information from industry that can lead to a joint development of a full flight-certified fission power system. Fission surface power technologies also will help NASA mature nuclear propulsion systems that rely on reactors to generate power. These systems could be used for deep space exploration missions.

NASA's fission surface power project is managed by the Glenn Research Center. The power system development is funded by the Space Technology Mission Directorate's Technology Demonstration Missions program, which is located at Marshall Space Flight Center.

5.3 LOCKHEED MARTIN TEAM

Lockheed Martin's space nuclear systems work includes three current contracts; a partnership with BWXT Technologies on both nuclear thermal reactor and fission surface power concepts for NASA and the Department of Energy, and a contract with the Defense Advanced Research Projects Agency to develop a spacecraft concept design with NTP capability.

While nuclear systems are an emerging field, Lockheed Martin has a long history and expertise in nuclear controls, having supported instrumentation and controls for both terrestrial power plants and Naval nuclear reactors. Lockheed Martin's expertise in avionics, mission control and integration give us leverage. The have also invested heavily in cryogenic hydrogen storage and transfer, as well as the overall nuclear reactor controls. The company also built the radioisotope thermoelectric generators for NASA planetary missions such as Viking, Pioneer, Voyager, Apollo, Cassini and New Horizons.

See Chapter 4.1.1 above for Lockheed Martin's work in this and other nuclear fields.

In addition to nuclear surface power, in 2021, Lockheed Martin was one of five companies selected to design a solar array concept that can autonomously deploy vertically and retract for relocation on the Moon. Also, in August 2022, Lockheed Martin was one of three companies selected to build the lunar solar array design into a prototype.

5.3.1 BWXT

See paragraph 4.1 above. BWXT is developing design concepts for a fission surface power system for NASA and the U.S. Department of Energy. FSP technologies are reliable, small and lightweight and could enable continuous power for deep exploration missions regardless of location, available sunlight and other natural environmental conditions.

5.3.2 Creare

Creare LLC is a leading innovator in the design and development of custom cryogenic components and systems, including vacuum systems, heat exchangers, control systems, and turbo-Brayton and other cryocoolers. They are located in Hanover, NH.

Creare develops systems with operating temperatures down to $4°$ K ($-452°$F) and cooling capacities from a few milliWatts to several kilowatts. The company has designed and developed cryocoolers for long-life space missions and low-cost terrestrial applications. It provides complete cryogenic services from conceptual design, analysis and optimization to hardware development, fabrication and testing.

Due to the large amounts of waste heat generated by these systems, a key consideration is the development of lightweight, highly efficient heat rejection systems (HRS) that can operate at elevated temperatures (~550 K). Currently, an approach that is being strongly considered is the use of titanium sheathed heat pipe with a carbon composite over-wrap, combined with a carbon composite radiator panel to decrease the system mass. Our innovation is the integration of an ultra-light radiator panel material with a lightweight titanium heat pipe. Our calculations show that our approach will reduce the total mass by as much as 20% compared to the carbon-composite systems under consideration and represents a lower risk approach to achieve a practical HRS.

5.4 WESTINGHOUSE TEAM

On June 29, 2022 NASA, in partnership with Battelle Energy Alliance, contractor for the U.S. Department of Energy's Idaho National Laboratory (INL), selected Westinghouse Electric Company to provide an initial design concept for a fission surface power system that could be ready to launch to the Moon by the end of the decade.

NASA and DOE selected three firms to prepare such design concepts to advance space nuclear technologies. The 12-month contracts are valued at US$5 million each and will fund the development of initial design concepts for a 40-kilowatt class fission power system planned to last at least 10 years in the lunar environment. Westinghouse Government Services LLC, a member of the Westinghouse family of companies, is partnered with Aerojet Rocketdyne and supported by Astrobotic for this effort.

"Westinghouse has a rich history of delivering innovative nuclear technology that provides safe, clean and reliable carbon-free power to even the most remote communities, so we are honored to help create groundbreaking technology for outer space," said David Durham, President, Westinghouse Energy Systems. "Westinghouse is committed to our partnership with Aerojet Rocketdyne and eager to begin our design for a fission surface power system."

5.4.1 Aerojet Rocketdyne
See also paragraph 4.2.2 above.
NASA is looking into reducing risk and increasing feasibility of nuclear propulsion for human-rated missions to Mars. In cooperation with NASA and other industry partners, Aerojet Rocketdyne has been leading the research effort on engine and mission architecture planning for this effort.

Nuclear thermal propulsion (NTP) has tremendous synergy with existing liquid rocket technologies; an area in which Aerojet Rocketdyne is a proven leader. This includes the development/production of turbomachinery and nozzles, cryogenic hydrogen fuel handling and providing heat management solutions.

Nuclear electric propulsion (NEP) relies on several key technologies, including high-power electric thrusters, power processing, power conversion, and power management and distribution, which are all areas in which Aerojet Rocketdyne is an established industry leader.

Aerojet Rocketdyne is working with manufacturers to leverage new low-enriched Uranium technology that is safer to handle and reduces expensive regulatory and security challenges; making both NTP and NEP attractive alternatives to comparable chemical and solar electric propulsion options for crewed deep space missions.

Aerojet Rocketdyne will support Westinghouse on their work on nuclear fission surface power efforts for the Moon and Mars.

5.5 IX JOINT TEAM

On June 21, 2022 the Department of Energy and NASA awarded IX, a joint venture between Intuitive Machines and X-energy, a $5 Million contract to conduct a one-year study to mature the design of a Fission Surface Power (FSP) solution that will deliver at least 40 kWe power flight system to the Moon by 2028.

Intuitive Machines' deep understanding of the Moon, lunar environments, and the company's ability to access and build complex spacecraft put Intuitive Machines in place to lead systems engineering and design for maturing an FSP solution. "This effective blend of companies brings existing and next-generation capabilities together to enable long-duration lunar surface missions," said Intuitive Machines President and CEO Steve Altemus. "Our capable team will provide an agile, affordable Fission Surface Power solution to further human and robotic exploration of the lunar surface."

X-energy is leading the development of an advanced small scale, portable reactor for space applications using its proprietary ceramic encapsulated fuel (TRISO-X) that can withstand temperatures four times greater than conventional nuclear fuel. Designed to be intrinsically safe, X-energy's space reactors deliver long life and high thermal power output at low mass based on numerous micro-reactor innovations developed over the past several years.

"Our highly capable team, including Maxar and Boeing, is extremely honored to be selected for the initial design of this U.S.-led lunar power project. We recognize the importance of a dependable power source that can maintain lunar and Mars habitats around the clock, especially during the 14-day lunar nights," said X-energy CEO Clay Sell. "This combined spaceflight-nuclear technology will protect the environment while

IX is a joint venture between Intuitive Machines and X-energy. Dr. Kam Ghaffarian is the founder and chairman of the board of both companies. Fission Surface Power is at the intersection of Intuitive Machines' and X-energy's strategic roadmap and leverages prior investment in reactor design, nuclear fuel, and lunar surface systems. IX assembled a best-of-industry team, combining capability in reactor design and operations, power conversion design, thermal management systems, and integrated space flight systems design with agility and innovative culture to deliver a complete Fission Surface Power solution.

5.5.1 Maxar

Maxar Technologies is a private company based in Westminster, CO. It specializes in manufacturing communication, Earth observation, radar, and on-orbit servicing satellites, satellite products, and related services. NASA recently awarded Maxar Technologies the first contract for its Lunar Gateway space station. It will build the power and propulsion element (PPE) of the planned gateway station. They were selected to be on the IX team.

5.5.2 Boeing

Boeing will be leveraging their ISS and satellite power generation and conversion systems experience.

IMAGE LINKS

Fig. 5.1 Concept of fission surface power on Mars.

https://www.nasa.gov/sites/default/files/styles/full_width/public/thumbnails/image/5power.jpg?itok=njpbjW1h

Fig. 5.2 Kilopower system annotated
https://www.researchgate.net/profile/Marc-Gibson/publication/326263176/figure/fig1/AS:674790817951745@1537894076132/Kilopower-1-kWe-nuclear-power-system-flight-concept-comparison-with-KRUSTY-nuclear-test_W640.jpg

Fig. 5.3 Marc Gibson demonstrating KRUSTY
https://www.spaceflightinsider.com/wp-content/uploads/2018/05/kilopower-michael-cole-15938-368x655.jpg

6

Technology

6.1 BASELINE CONSIDERATIONS

Advanced nuclear propulsion systems (alone or in combination with chemical propulsion systems) have the potential to substantially reduce trip time compared to fully nonnuclear approaches. The National Academy of Science and NASA have assessed the primary technical and programmatic challenges, merits, and risks for developing nuclear thermal propulsion (NTP) system or nuclear electric propulsion (NEP) system augmented with a chemical propulsion system for the human exploration of Mars.

Many NASA studies have considered the use of NTP or NEP to facilitate the human exploration of Mars. Mission scenarios associated with nuclear, solar, and chemical propulsion systems and various mission parameters were studied. Launch assumptions varied with the launch systems in use or under development at the time of each study. Because crewed Mars missions are significantly more challenging in terms of launch mass and trip time than all prior space missions, in-space propulsion is a critical technology. This is evident by the wide range of propulsion systems that have been considered over multiple mission studies.

Based on the relative orbits of Mars and Earth, the distance between Earth and Mars ranges from 55 to 400 million km (34 to 249 million mi) over a synodic period of approximately 26 months. Launch (or Earth departure) requirements vary significantly over this cycle. Each 26-month cycle is not the same. Propulsion system performance requirements, in terms of the total velocity increment (ΔV) of a round-trip Mars mission, vary from one launch opportunity to the next. The ΔV for a particular mission also depends on other mission constraints, particularly the stay time at Mars and the desired trip time.

There are two classes of crewed missions to Mars: conjunction class and opposition class. Conjunction-class missions have the lowest ΔV requirements. For crewed conjunction-class missions, trip times are typically 180 to 210 days each way, stay times on Mars are typically 500 days or more, and total mission time is around 900 days. These are called the "long stay" missions. In contrast, one leg of opposition-class missions occurs when the orbital alignment of Earth and Mars is less favorable, but they allow for "short stays" on the surface of Mars. These missions have higher ΔV requirements and require more propellant, which increases the mass of the Mars vehicle and the number of launch vehicles necessary to lift the required mass to its assembly orbit.

Opposition-class missions are characterized by much shorter stay time on Mars (30 to 90 days) and a shorter total mission time (400 to 750 days). An additional complexity of opposition-class missions is that the long leg of the mission typically passes inside Earth's orbit, generally as close to the Sun as the orbit of Venus, to mitigate the adverse planetary alignment of that leg of the mission. This results in both thermal and radiation challenges for a crewed Mars mission.

The baseline mission specified by NASA is an opposition-class crewed mission to Mars launched in 2039. This mission would be preceded by cargo missions beginning in 2033 to pre-place surface infrastructure and consumables for the crew. The propulsion system needed for this mission would also be sufficient for conjunction-class missions. The baseline mission has the following parameters:

- Crew mission launch in 2039 opportunity;
- Total crew trip time ≤750 days;
- Split mission with separate crew and cargo vehicles,
  Same propulsion systems used on all vehicles,
  Cargo vehicles arrive at Mars prior to first crew departure from Earth;
- Stay time on the Mars surface of 30 days;
- Crew of four, two of whom land on Mars; and

- Vehicle systems, cargo, and propellant launched by multiple launch vehicles to an assembly orbit, which would be either in low Earth orbit or cislunar space.

In order to meet the requirement for total trip time with an NEP system, Earth departure and Mars capture and departure would be augmented by an additional in-space liquid methane and liquid oxygen chemical propulsion system. The NEP system provides acceleration and deceleration in interplanetary space. In contrast, the NTP system provides propulsion for all transit maneuvers.

As Earth and Mars revolve about the Sun, the most efficient trajectories vary, resulting in varying levels of propulsive requirements (ΔV) over a 15- to 17-year period. A factor in mission assessment for repeated trips to Mars is the ability of propulsion systems to meet mission ΔV requirements over a series of consecutive launch opportunities without large variability in overall mission parameters, such as propellant mass, which could drive very different launch requirements for different opportunities. This variability is reduced by propulsion systems with high specific impulse (Isp). Previous studies have shown the impact of NTP for an opposition-class mission in different launch opportunities, although not for the current years of interest. For example, the change in vehicle (propellant) mass with launch date for an advanced chemical system with an Isp of 480 s as compared to a NTP system vehicle with an Isp of 825 s is most significant. The mass variation with launch opportunity for the higher Isp system is about one half of the variation of the chemical system.

Similar benefits would likely be achieved with an NEP system with an Isp of 2,000 s paired with a conventional chemical system. This is particularly important because some launch opportunities are not feasible using purely a chemical system. Flexibility in launch date is a major architectural advantage of the use of nuclear propulsion.

With these baseline factors in consideration, the following are those nuclear technologies being considered for lunar and Mars surface applications and deep space propulsion.

6.2 NUCLEAR ELECTRIC PROPULSION

It's too soon to say what propulsion system will take the first astronauts to Mars, as there remains significant development required for each approach. We know it needs to be nuclear-enabled to reduce travel time. NASA is advancing multiple options, including nuclear electric and nuclear thermal propulsion. Both use nuclear fission but are very different from each other. A nuclear electric rocket is more efficient, but it doesn't generate a lot of thrust. Nuclear thermal propulsion, on the other hand, provides much more thrust. Whichever system is selected, the fundamentals of nuclear propulsion will reduce the crew's time away from Earth. The agency and its partners are developing, testing, and maturing critical components of various propulsion technologies to reduce the risk of the first human mission to Mars.

Nuclear electric propulsion systems use propellants much more efficiently than chemical rockets but provide a low amount of thrust. They use a reactor to generate electricity that positively charges gas propellants like xenon or krypton, pushing the ions out through a thruster, which drives the spacecraft forward. Using low thrust efficiently, nuclear electric propulsion systems accelerate spacecraft for extended periods and can propel a Mars mission for a fraction of the propellant of high thrust systems.

NASA's Marshall Space Flight Center leads the agency's space nuclear propulsion project in partnership with a DOE team that includes scientists and engineers from Idaho National Laboratory, Los Alamos National Laboratory, and Oak Ridge National Laboratory. STMD's Technology Demonstration Missions program funds the technology development.

Nuclear electric propulsion builds on NASA's work maturing solar electric propulsion thrusters and systems for Artemis, as well as the development of fission power for the lunar surface. Significant investment has also been made in relevant fuel and reactor technologies for small, terrestrial reactors that could be adapted to space reactors to power electric propulsion. The U.S. government's aim to establish a fuel fabrication capability has a range of applications, including nuclear propulsion and fission surface power.

6.2.1 System Concept

The following description is courtesy of the National Academies of Sciences. Nuclear electric propulsion (NEP) systems convert heat from the fission reactor to electrical power, much like nuclear power plants on Earth. This electrical power is then used to produce thrust through the acceleration of an ionized propellant. An NEP system can be defined in terms of six subsystems, which are shown in the figure and briefly described below:

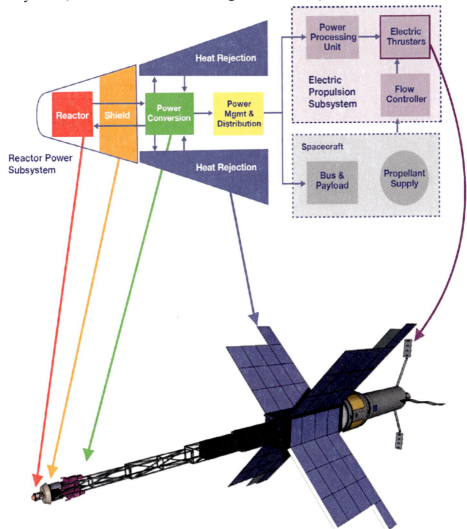

Fig. 6.1 NEP Subsystems. Graphic courtesy of Lee Mason, NASA.

1. Reactor. As with a nuclear thermal propulsion (NTP) system, the reactor subsystem produces thermal energy. In an NEP system, this thermal energy is transported from the reactor to the power conversion subsystem through a fluid loop.

2. Shield. As with an NTP system, the shield subsystem reduces the exposure of people and materials in the vicinity of the reactor to radiation.

3. Power conversion. The power conversion subsystem converts some of the thermal energy transported from the reactor to electrical energy through either dynamic mechanical or static solid-state processes. For example, flow a heated fluid through turbines or through the use of semiconductor or plasma diodes to move charged particles through a material. The remaining thermal energy is rejected as waste heat.

4. Heat rejection. Terrestrial power systems can use ambient water and air for convective cooling. The thermal energy created by NTP systems is transferred to the cryogenic propellant and exhausted into space. High power NEP systems require heat rejection radiators with large surface areas to provide adequate cooling, and, as power levels increase, the size and mass of the heat rejection subsystem has the potential to dominate over other subsystems. Heat rejection at high temperatures reduces the radiator area since radiation increases proportionally to the fourth power of the absolute temperature of the radiator. High-temperature operation thereby increases performance, but it becomes a challenge for other aspects of the system.

5. Power management and distribution (PMAD). Electrical power from the power conversion subsystem is often generated near the reactor to avoid thermal losses; however, the power must be controlled and distributed over relatively large distances to the electric propulsion (EP) subsystems. The PMAD subsystem consists of the electronics, switching, and cabling to manage the electrical voltage, current, and frequency of the transfer efficiently

6. The EP subsystem converts electricity from the PMAD subsystem into thrust through electrostatic or electromagnetic forces acting on an ionized propellant. The EP subsystem consists of the power processing unit (PPU), propellant management system (PMS), and thrusters. The PPU converts the power provided by the PMAD to a form that can be used to generate and accelerate a plasma. A "direct-drive" system would directly drive the EP subsystem from the PMAD subsystem with a commensurate reduction in PPU mass. Power control hardware for switching and power quality would still be required for starting, throttling, and managing transients and faults within the EP subsystem. The PMS manages the propellant flow to the thrusters.

NEP system performance is governed by the total system mass required to produce the required power level i.e., the system specific mass, in kilograms per kilowatt-electric (kg/kWe), the performance of the EP subsystem, and the lifetime and reliability of all subsystems. System design trades focus on maximizing the power conversion subsystem efficiency, the waste heat rejection temperature, and the efficiency and specific impulse (Isp) of the EP subsystem while achieving the mission lifetime and reliability requirements.

6.2.2 State-of-the-Art

An integrated technology development program aimed specifically toward an NEP system operating at more than 1 MWe has not been undertaken. Although preliminary design studies for MWe-class NEP systems have been conducted, there have not been any significant detailed design, hardware development, or modeling and simulation (M&S) advances for the full, integrated NEP system. NEP technologies, designs, and M&S tools related to HEU fuels, power conversion, heat rejection, and thrusters have been developed for 100 to 200 kWe NEP systems; some of these technologies could be scaled to the megawatt electric power level. Developing an NEP system for the baseline mission will likely involve the use of multiple NEP modules which, in the aggregate, will provide the total propulsive power. This would increase system complexity, especially since the NEP system design includes six major subsystems (on each NEP module), and the spacecraft would also need to incorporate a chemical in-space propulsion system.

6.2.3 Thrusters

Electric Propulsion (EP) systems have been used for spaceflight for decades, but to date the available power level has been limited to kilowatt-electric, not megawatt electric, and the source of power has been solar panels. Of the various thruster types that have been used, the two most likely to provide the required performance and lifetime capabilities for Mars missions at the required power levels are ion thrusters and Hall thrusters. Both of these types of thrusters have extensive flight heritage at power levels below 5 kWe.

Ion thrusters use two or more parallel grids with a voltage applied to each to extract and accelerate ions created in a discharge chamber upstream of the grids. Because ions are extracted and accelerated through the grids, a cathode neutralizer is needed to emit electrons to prevent a charge imbalance from forming. Charge separation in the grid assembly limits the maximum thrust density of ion thrusters, meaning that 100 kWe class ion thrusters are likely quite large. Ion thruster modeling and simulation (M&S) is well developed, with good predictive performance and lifetime models that will support scaling to 100 kWe class thrusters. The primary area of uncertainty in ion thruster M&S is the impact of ground test facilities on long-duration thruster life tests.

For a 3:30 min video about the Dawn ion thruster narrated by Dr. Marc Rayman, go to:
https://cdn.jwplayer.com/previews/7b87xVE1

For a 2:45 minute video about the NEXT-C xenon thruster on DART, go to:
https://youtu.be/aLTAjDZ49vo

The NEXT engine is a type of electric propulsion in which thruster systems use electricity to accelerate the xenon propellant to speeds of up to 145,000km/hr or 40 km/s (90,000mph). NEXT can produce 6.9 kW thruster power and 236 mN of thrust. It can be throttled down to 0.5kW power, and has a specific impulse of 4,190 seconds (compared to 3,120 for NSTAR).

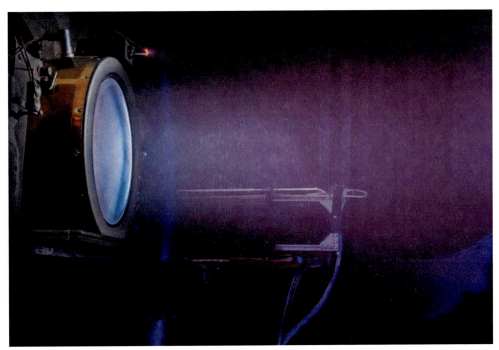

Fig. 6.2 The ion thruster used on DART. Photo courtesy of NASA.

For more on the DART mission and its use of ion thrusters, read the book; *Asteroid Detection and Mitigation in the 21st Century* available at Amazon.

With Hall thrusters, propellant is injected through an annular channel and ionized by electrons trapped by an applied radial magnetic field. A voltage difference is applied between the anode, which usually serves as the propellant injector at the upstream end of the channel, and a downstream hollow cathode that supplies the electrons to the channel. The mixture of electrons and ions in the acceleration zone means that the thruster does not have the thrust density limitation associated with ion thrusters, although other lifetime considerations limit the achievable thrust densities. As with ion thrusters, M&S tools for Hall thrusters are well advanced and will support scaling to 100 kWe thrusters, although ground testing of high-power Hall thrusters has revealed that interactions among the test facility, the thruster, and its conducting plasma plume can impact the performance and lifetime measurements in ways that are not fully understood. This introduces uncertainty into current predictions of in-space performance and lifetime for high-power Hall thrusters.

Fig. 6.3 A Hall Thruster. Photo courtesy of NASA/GRC.

Both ion and Hall thrusters have flown in space; for example:

- The Aerojet Rocketdyne XR-5 Hall thruster, which is currently in use on several DOD and commercial spacecraft and has been ground tested to more than 10,000 hr.
- NASA's Advanced Electric Propulsion System (AEPS) Hall thruster, which is undergoing flight development, has a projected lifetime of more than 20,000 h and is slated for NASA's Lunar Gateway Power and Propulsion Module
- NSTAR ion thruster, which flew on Deep Space 1 (1998) and DAWN (2007), was life tested to more than 30,000 hours.
- NASA's Evolutionary Xenon Thruster–Commercial (NEXT-C) thruster, which was ground tested for 50,000 hours and was used on the Double Asteroid Redirection Test (DART) mission (2021).

All flight thrusters also have flight Power Processing Units (PPU) and power management system (PMS) subsystems, although they are designed to interface with a solar photovoltaic power system, not a nuclear power source.

6.2.4 Summary

At a concept modeling and analysis level, NEP shows promise for the baseline mission. However, intermittent funding has resulted in very limited, if any, advance in its technology readiness since 2005, and that work focused on 200 kWe NEP systems, not the MWe-class system required for this application. The need to extrapolate from those results to a 1 to 2 MWe system required for the baseline mission without increasing specific mass results in considerable uncertainty in feasibility of this path on a timeline consistent with the baseline mission. In particular, uncertainty in fuel system architecture and the significant scaling of thruster requirements and thermal and power management are considerable challenges. The reliability and lifetime requirements of such a system merit careful attention and the lack of any substantive integrated system test remains a challenge. The present state of NEP technology and limited subsystem ground test facilities for reactors and high-power EP thrusters require near-term assessment. Advanced reactor test facilities are currently under development for terrestrial programs, but the extent to which those facilities would be able to contribute to the development of MWe-class NEP systems remains to be determined.

EP has benefited from gradual increases in power level for solar powered spacecraft. There are currently hundreds of kilowatt-electric-class spacecraft flying operationally, and a 40 kWe SEP system, using multiple 13 kWe thrusters, is projected to launch in 2024. However, testing thrusters at power levels above 50 kWe, particularly for in-space performance and lifetime, will challenge existing vacuum facility capabilities.

In 2021, the National Academies of Science stated that if NASA plans to apply nuclear electric propulsion (NEP) technology to a 2039 launch of the baseline mission, NASA should immediately accelerate NEP technology development.

6.3 NUCLEAR THERMAL PROPULSION

Nuclear thermal propulsion has been on NASA's radar for more than 60 years. See Chapter 2. The new hardware design and development phase pursued through a request for proposals released February 12, 2021, builds on existing efforts to mature crucial elements of a nuclear thermal propulsion system.

NASA, in partnership with DOE, is developing and testing new fuels that use low-enriched uranium for space applications to see how they perform under the extreme thermal and radiation environments needed for nuclear thermal propulsion. NASA is working closely with DOE, industry, and universities to put fuel samples in research reactors at the Idaho National Laboratory's Transient Reactor Test (TREAT) facility and the MIT Nuclear Reactor Laboratory for nuclear testing. The team is also performing non-nuclear testing in simulated reactors at Marshall test facilities.

"The reactor underpinning a nuclear thermal propulsion system is a significant technical challenge due to the very high operating temperatures needed to meet the propulsion performance goals," explained Anthony Calomino, NASA's nuclear technology portfolio lead within STMD. While most of the engine operates at modest temperatures, materials in direct contact with the reactor fuel must be able to survive temperatures above 4,600° F. NASA and DOE have been working with industry on a viable approach, and industry will now develop preliminary designs to meet this challenge.

"We're exploring both nuclear electric and nuclear thermal propulsion options for crewed Mars missions," Calomino said. "Each technology has its unique advantages and challenges that need to be carefully considered when determining the final preference." Whichever propulsion system is ultimately chosen, the fundamentals of nuclear propulsion can enable robust and efficient exploration beyond the Moon. NASA will continue to develop, test, and mature various propulsion technologies to reduce risk and inform the Mars transport architecture.

6.3.1 System Concept
A nuclear thermal propulsion (NTP) system is conceptually similar to a chemical propulsion system, where the combustion chamber has been replaced

by a nuclear reactor to heat the propellant. Chapter 2 discussed the basic concepts and included figures of previous nuclear rocket programs. Fig. 2.6 showing the NERVA diagram shows the basic concept. Although the Rover/NERVA programs demonstrated proof of concept for an NTP system, the programs were cancelled before program goals were achieved due to a shift in funding priorities. Consequently, no complete NTP system has been assembled and tested in its flight configuration or flown in space. Other NTP programs have been carried out since Rover/NERVA, but none have built any additional reactors or engines.

6.3.2 State-of-the-Art

As of the National Academies of Sciences report in 2021, the current state of the art for the reactor subsystem was limited to the modeling and simulation (M&S) capabilities used to analyze a reactor virtually. At that time, existing hardware manufacturing capabilities were insufficient to build an NTP reactor at the scale required for cargo or crewed missions associated with the baseline mission. Since that time, DARPA, DOD laboratories and NASA have let contracts to allow industry to advance the state-of-the-art with their input to the design of nuclear thermal propulsion reactors and systems.

6.3.3 Reactors

An NTP system with a propellant reactor exit temperature of approximately 2700° K (4400 ° F) represents an extreme environment in terms of temperature and hydrogen corrosion for the materials in the reactor core. This reactor operating temperature implies that there are few viable fuel architectures. The fuel element, which includes the fuel and cladding, the fuel assemblies, moderator, support structures, and the reactor pressure vessel must maintain physical integrity while cycled through the thermomechanical stress induced during repeated cycles of reactor startup, operation at power, shutdown, and restart.

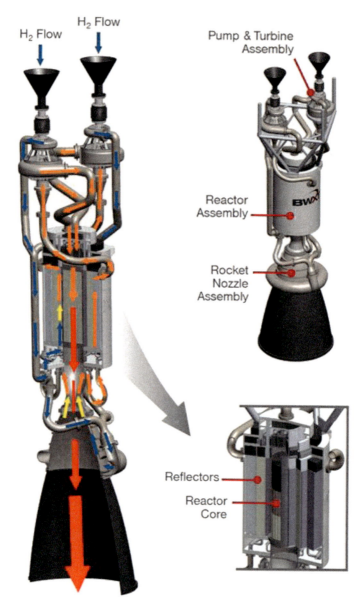

Fig. 6.4 A nuclear thermal propulsion reactor. Diagram courtesy of BWXT

Nuclear thermal propulsion engine diagram with arrows that show the flow path of the hydrogen propellant. Blue arrows represent the coldest relative gas temperatures and red arrows represent the hottest. Credit / BWX Technologies

6.3.4 Support Systems

Engine hardware, such as turbomachinery and valves, has evolved independent of NTP reactor hardware for use on chemical propulsion systems. Existing chemical propulsion engine components can be scaled, modeled, and integrated for NTP use. For instance, the RL-10 and similar turbopumps have been modeled for decades for a variety of NTP design studies while having undergone maturation and hardware testing for a variety of chemical propulsion uses, including in space.

M&S capabilities applicable to the nonnuclear (i.e., chemical) engine subsystem elements are well developed for both static and dynamic engine flow conditions. Some of these models are also applicable to NTP engine subsystems and have been used to model both HEU and HALEU-type engines.

Long-term storage and active cryogenic technologies for LH2 have similarly evolved independently of NTP, but significant challenges must still be overcome to meet a storage time of perhaps 4 years for the baseline mission (2 years in assembly plus 2 years for the round trip to Mars). Ongoing research technology development by NASA will lead to several missions to demonstrate advanced technologies for the storage and transfer of cryogenic fluids in space.

6.3.5 Cryogenic Hydrogen Storage

The development of multiyear cryogenic storage capabilities for LH2 remains a significant challenge for NTP systems. Storage of metric tons of LH2 at cryogenic temperatures as low as 20 K (-424°F) with minimal losses, is needed because of the long duration of the baseline mission, including time for in-space vehicle assembly and the round trip to Mars.

The current expectation for the baseline mission is that at least six NTP system starts will be needed, with a total LH2 propellant requirement that ranges from 7 to 21 10,000-kg (11 tons) tanks of LH2 depending on which launch vehicles are used and the mission departure year. Minimizing the boiloff of LH2 from the storage tanks is necessary to reduce the amount of LH2 that must be launched and the number of storage tanks that must be integrated into the Mars exploration spacecraft.

Although development of refrigeration technology is proceeding, existing cryocooling systems cannot reliably meet propellant tank requirements over a mission of this duration. Additionally, propellant mass must be accurately measured before and after each firing of the propulsion system to appropriately balance the flow rate to the reactor startup and reactivity control operations. Cryocooling systems will require electrical power throughout the mission, which would be provided by small solar arrays that are dedicated to this purpose.

Note that NASA MSFC and GRC have been working on the eCryo project for many years.

6.4 HIGH-ASSAY LOW- ENRICHED URANIUM (HALEU)

6.4.1 How it is Manufactured

More than 20 U.S. companies are developing advanced reactors. Most of these new reactor designs will be smaller, more flexible and less expensive to build and operate. The majority of these designs will require a fuel that isn't yet available at a commercial scale. It's what the industry calls High-Assay Low-Enriched Uranium, or HALEU for short. Some of these companies are depending on it.

To know what this fuel is; is to know what it is not. Our national fleet of power reactors runs on uranium fuel that is enriched up to 5% with uranium-235; the main fissile isotope that produces energy during a chain reaction. This is referred to as Low Enriched Uranium (LEU). In comparison, nuclear submarine reactors run on 95% uranium; referred to as High Enriched Uranium (HEU). Older atomic bombs used 90% or greater HEU. By definition, High Assay Low-Enriched Uranium (HALEU) is enriched between 5% and 20% and is required for smaller U.S. advanced reactors designs that get more power per unit of volume. HALEU will also allow developers to optimize their systems for longer life cores, increased efficiencies and better fuel utilization.

The DOE projects that more than 40 metric tons of HALEU will be needed before the end of the decade, with additional amounts required each year, to deploy a new fleet of advanced reactors. To help mitigate that risk, DOE is exploring three options to support the testing and demonstration of these advanced reactors with HALEU fuel.

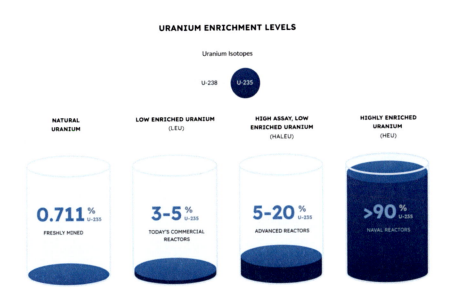

Fig 6.5 Various Uranium Enrichment Levels. Photo courtesy of Centrus.

The DOE and its national laboratories are working on two chemical processes to provide small amounts of HALEU to vendors in the near-future. Both methods involve the recycling of used nuclear fuel from government-owned research reactors to recover highly enriched uranium (greater than 20%) that can then be down-blended to make HALEU fuel.

Irradiated fuel from DOE-research reactors is prepared and placed into a high-temperature molten salt chemical bath. An electric current is then used to separate the highly enriched uranium metal from the fission products. The recovered uranium is cleaned and mixed with lower enriched uranium to create HALEU. The uranium is then fabricated into new fuel in a high-temperature furnace. The Idaho National Laboratory (INL) is working to make up to 10 metric tons of HALEU using this process in the near-term to support current testing and demonstration projects.

Irradiated fuels are dissolved in hydrochloric acid gas to remove the aluminum or zirconium cladding. The fuel is then passed through a modular solvent extraction system to separate the uranium from its fission products. The uranium is then down-blended with lower enriched uranium and returned to its solid form to produce HALEU.

INL is currently testing a small-scale pilot facility on unirradiated materials to research and scale-up a new Hybrid Zirconium Extraction Process (ZIRCEX) process. Argonne, Oak Ridge and Pacific Northwest national laboratories are collaborating on this project. A three-year demonstration project is underway to send a strong signal to potential vendors that there will be a proven domestic capability to produce HALEU when the market demands it.

6.4.2 Centrus centrifuge cascade

DOE has partnered with Centrus Energy Corporation of Bethesda, MD and its wholly-owned subsidiary, American Centrifuge Operating LLC to pioneer production of High-Assay, Low-Enriched Uranium (HALEU) and to manufacture 16 advanced centrifuges for deployment at an enrichment facility in Piketon, Ohio. The company's AC-100M machine was developed through the years with support from DOE and will demonstrate enrichment of uranium hexafluoride gas to produce HALEU. It will then be used for advanced reactor fuel qualification testing and reactor demonstration projects. The AC-100M technology will be available for commercial deployment at the conclusion of the demonstration.

Under a previous cost-shared contract signed with the DOE in 2019, Centrus installed centrifuges and secured a license amendment from the Nuclear Regulatory Commission (NRC) in June 2021 to enrich uranium up to 20 percent fissile U-235. In November 2022, the DOE announced a new contract with Centrus to finish the cascade, complete final regulatory steps, begin operating the cascade, and produce HALEU.

Specifically, the new contract called for Centrus to enrich uranium hexafluoride gas to produce 20 kg (44 lbs) of 19.75 % enriched HALEU by December 3, 2023. An annual production rate of 900 kg (1984 lbs) of HALEU per year would be expected thereafter, subject to appropriations, with options to produce more material under the contract in future years.

The DOE is exploring different avenues to meet domestic needs through its HALEU Availability Program, with a focus on a sustainable commercial supply chain. That includes using $700 million of earmarked funds from the Inflation Reduction Act to initiate a program of offtake contracts to stock a DOE-owned HALEU bank. Industry projections vary, but the DOE projects that more than 40 metric tons of HALEU will be needed before the end of the decade, with additional amounts required each year.

On February 9, 2023, Centrus Energy announced that it has finished assembling a cascade of uranium enrichment centrifuges and most of the associated support systems ahead of its contracted demonstration of high-assay low-enriched uranium (HALEU) production by the end of 2023. When the 16-machine cascade begins operating inside the Piketon, Ohio, American Centrifuge Plant, which has room for 11,520 machines, it will be the first new U.S. technology based enrichment plant to begin production in 70 years.

Fig. 6.6 Centrus uranium enrichment cascade. Photo courtesy of Centrus/ American Centrifuge Operating LLC.

Centrus has said that a full-scale HALEU cascade, consisting of 120 individual centrifuge machines, could be brought on-line about three and a half years after funding is secured, and that subsequent machines could be added every six months. A 120-machine cascade could produce about six metric tons of HALEU per year.

Centrus has completed initial testing of the cascade and support systems. Before operations can begin, Centrus must finish construction of the remaining support systems, including a fissile materials storage area. Final NRC operational readiness reviews will be required before cascade operations and HALEU production can begin by the end of 2023.

6.5 TRISO

A Tri-Structural Isotropic (TRISO) particle is a Uranium bearing sphere coated with special ceramic layers designed like tiny pressure vessels. The layers contain fission products inside and ensure mechanical and chemical stability during irradiation and temperature changes. Coated particle fuel was first developed for the Dragon Reactor in the UK in the 1960s, and follow-on gas reactors in Germany, the United States, Japan, South Africa, and China. The fuel is extremely robust, with over 60s years of development and experience.

X-energy LLC, (See Chapter 4.2.1) is one of the only companies in the U.S. capable of producing ceramic-coated fuel forms using HALEU, which is at the core of TRISO fuel. Each TRISO fuel kernel consists of a 0.5 micron pellet of uranium oxycarbide; the size of a poppy seed, wrapped in three alternating layers of graphite and silicon carbide. Thousands of these particles are embedded in a graphite fuel form; either pebbles or prismatic compacts. In X-energy's terrestrial reactor, the Xe-100, more than 60,000 of these pebbles (roughly the size of a cue ball) will be cycled through the reactor core over the course of a year. This concept can be customized for many applications, including powering lunar and Mars bases.

Fully Ceramic Microencapsulated (FCM®) Fuel is a new approach to inherent reactor safety by providing an ultimately safe fuel. Industry standard TRISO fuel, which contains the radioactive byproducts of fission within layered ceramic coatings, is encased within a fully dense silicon carbide matrix. This combination provides an extremely rugged and stable fuel with extraordinary high temperature stability.

To make FCM Fuel, X-energy starts by additively manufacturing SiC shells in whatever shape needed; usually cylinders or annular cylinders. TRISO particles are then poured into the shell. Finally, SiC is deposited into the shell and between the particles to form a dense and robust matrix. The manufacturing process allows for a wide variety of fuel shapes beyond simple pins including annular cylinders, hollow pebbles, Y-compacts and many other particular specifications.

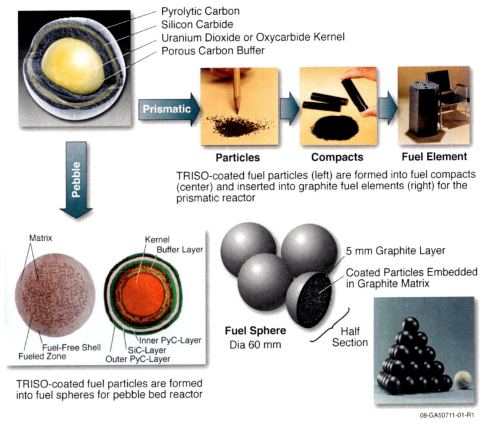

Fig. 6.7 Types of TRISSO fuel. Photo courtesy of DOE INL.

X-energy has been working closely with the DOE to safely and reliably manufacture, test, and qualify TRISO-X fuel for years. This has laid the foundation for fueling our high temperature gas reactors on Earth and beyond. "Close collaboration with our TRISO-X fuel team and our sister companies who have been planning space missions for years has greatly accelerated our multi-disciplinary design," says Dr. Brad Rearden, the Director of Engineering and architect for space reactors. "We are leveraging lessons learned from other micro-reactor initiatives and leaning on our custom advanced computational toolkit to create a truly innovative solution that can be manufactured, tested, and flown on NASA's accelerated timeline."

NASA's fission surface power (FSP) program requires a company to rapidly produce a functioning reactor for ground testing prior to producing a unit for spaceflight. The aim is to have a fully capable nuclear power station ready to launch to the Moon by late 2026. Designs must also be extendable for human habitats on the surface of Mars.

Note that on August 18, 2022 Ultra Safe Nuclear Corporation , a global leader in the development of micro reactors, today announced the opening of its Pilot Fuel Manufacturing (PFM) facility in Oak Ridge, TN located in the East Tennessee Technology Park (ETTP). The facility is leveraging the region's specialized workforce to produce the first fuel for testing and qualification for use in Ultra Safe Nuclear Corporation's advanced Micro Modular Reactor (MMR®) Energy System. The New pilot manufacturing facility will produce TRISO and FCM® fuel for qualification for U.S. advanced micro reactors. The production-scale radiological manufacturing line includes advanced manufacturing process licensed from ORNL.

6.6 FISSION SURFACE POWER

NASA's Glenn Research Center has been working on fission surface power for many years. Initially, this capability will be provided on the Moon for the Artemis Program. The first lunar landing is planned for 2025 so the capability must be available soon. To ensure that these missions have all the energy they need to conduct operations, NASA is developing a Fission Surface Power (FSP) system that will provide a safe, efficient, and reliable electricity supply.

In conjunction with solar cells, batteries, and fuel cells, this technology will allow for long-term missions to the Moon and Mars in the near future.

Over decades, the Artemis program will evolve into a sustained presence on the Moon in the form of an Artemis Base Camp. See the book; *Artemis Base Camp; The First Steps* available on Amazon. Having fission reactors for lunar surface operations is vital to establishing a "sustained lunar exploration." capability. This technology provides the capability of achieving self-sufficiency in terms of resources, materials, and energy.

"This base will require a considerable amount of electricity so astronauts can recharge rovers, conduct experiments, and produce water, propellant, building materials, and oxygen gas using the Moon's natural resources; a process known as In-Situ Resource Utilization (ISRU)", said Jim Reuter, the Associate Administrator for NASA's Space Technology Mission Directorate (STMD), which funds the fission surface power project.

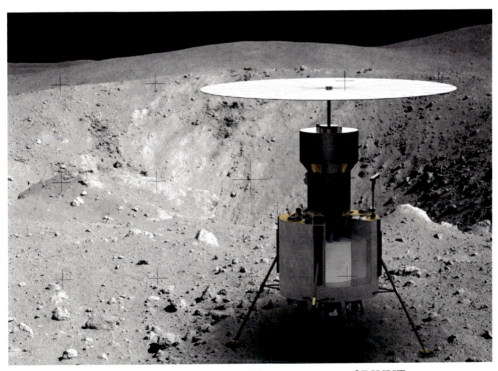

Fig. 6.8 Lunar Fission Surface Power. Photo courtesy of BWXT.

"Plentiful energy will be key to future space exploration," he said in a NASA press release. "I expect fission surface power systems to greatly benefit our plans for power architectures for the Moon and Mars and even drive innovation for uses here on Earth." The concept builds on NASA's Kilopower project, an effort to create a small, lightweight fission system that could provide up to 10 kilowatts (kW) of power continuously for at least ten years.

The Kilopower project started at the NASA Glenn Research Center in 2015 and went through design, development and test; completing the project in March, 2018 with the successful completion of the Kilopower Reactor Using Stirling Technology (KRUSTY) demonstrator. This prototype consisted of a solid, cast uranium-235 reactor core (about the size of a paper towel roll) and passive sodium heat pipes that transferred the heat generated by slow fission reactions to high-efficiency Stirling engines, which convert the heat to electricity. See Chapter 5.

Based on this success, NASA has since partnered with the DOE's Idaho National Laboratory (INL) operated by Battelle Energy Alliance to develop the Kilopower-inspired FSP for the Artemis Program. This will culminate with a technology demonstration, tentatively scheduled for the early 2030s, which will see a prototype reactor be sent to the Moon to test its capabilities under lunar conditions.

For a 3:09 minute video on Kilopower, go to:
https://youtu.be/fugONNLb9JE

As Todd Tofil, the FSP Project Manager at NASA's Glenn Research Center explained: "NASA and the DOE are collaborating on this important and challenging development that, once completed, will be an incredible step towards long-term human exploration of the Moon and Mars. We'll take advantage of the unique capabilities of the government and private industry to provide reliable, continuous power that is independent of the lunar location."

On November 19th, 2021 NASA and the DOE issued a solicitation to American companies for design concepts for a fission power system that could be ready to launch within a decade for a demonstration on the Moon. As stated in solicitation documents, the FSP should consist of "a uranium-fueled reactor core, a power conversion system (PCS), a thermal management system, and a power management and distribution (PMAD) system having a capability to provide no less than 40 kilowatts of continuous electric power at the user interface at end of life.

In conjunction with conventional methods, compact and lightweight fission reactors are considered ideal for providing electricity for lunar exploration. For starters, fission systems are reliable, capable of operating continuously in the permanently-shadowed craters that dot the Moon's South Pole-Aitken Basin. During lunar nights (which last 14 Earth days), solar power is largely unavailable, making reactors highly desirable. Because the technology has matured over many years, nuclear power systems can also be scaled to create compact and lightweight systems. This is key to fulfilling NASA's Artemis mission requirement for power systems capable of autonomous operation from the deck of a lunar lander or a lunar surface rover.

In 2022, NASA and the DOE selected competing U.S. companies to develop initial designs. See Chapter 5 for the contracting details.

6.7 MODERATORS

Neutron moderators are a type of material in a nuclear reactor that work to slow down the fast neutrons (produced by splitting atoms in fissile compounds like uranium-235), to make them more effective in the fission chain reaction. This slowing or moderation of the neutrons allows them to be more easily absorbed by fissile nuclei, creating more fission events.

Materials used for moderation need to have a very specific set of properties. First, a moderator cannot absorb neutrons itself. This means that the moderator should have a low neutron absorption cross-section. However, the moderator should be able to slow down neutrons to an acceptable speed. Thus, in an ideal moderator, the neutron scattering cross-section is high. This neutron scattering is a measure of how likely a neutron will interact with an atom of the moderator. If the collisions between neutrons and nuclei are elastic collisions, it implies that the closer in size the nucleus of an atom is to a neutron, the more the neutron will be slowed. For this reason, lighter elements tend to be more efficient moderators. Common moderators have a low neutron absorption cross-section but a comparatively large neutron scattering cross-section.

6.7.1 Types of Moderating Materials

There are several different types of moderating materials, and each have places where they are used more effectively. Typically-used moderator materials include heavy water, light water, and graphite. The relative properties of these materials are compared below. The moderators vary in terms of their moderating abilities, as well as in their costs.

Light water

Light water (no different than regular water) is used in many reactors because it contains large amounts of hydrogen. Hydrogen works well as a neutron moderator because its mass is almost identical to that of a neutron. This means that one collision will significantly reduce the speed of the neutron because of the laws of conservation of energy and momentum. In addition, light water is abundant and fairly inexpensive. One drawback is that hydrogen has a relatively high neutron absorption cross-section because of its ability to form deuterium. Thus light water can only be used as a moderator along with enriched fuels. Reactors that use light water are known as light water reactors and include the pressurized water reactor (PWR), the boiling water reactor (BWR), and the supercritical water cooled reactor (SCWR).

Heavy water

Heavy water is used in reactors because its benefits are similar to light water, but since it contains deuterium atoms, its neutron absorption cross section is much lower. The main disadvantage to the use of heavy water is its high cost of production, as it is made using the Girder-Sulfide process. Reactors that use heavy water include the CANDU designs and the pressurized heavy water reactor.

Graphite

Graphite has been a popular moderator in the past; however, one drawback is that it needs to be extremely pure to be effective. Graphite can be made artificially using boron electrodes, however, since boron is a very good neutron absorber; a small amount of contamination will make the graphite an ineffective moderator. One benefit of graphite is that even at the high purity that is necessary for graphite to perform well; it is available at a fairly low price. In addition, graphite is a good moderator as it is thermally stable and conducts heat well. However, at high temperatures the graphite can react with oxygen and carbon dioxide in the reactor and this decreases its effectiveness. Another potential issue with using graphite as a moderator is its ability to oxidize in the presence of air, and its low strength and density which could cause it to change dimensions in the reactor.

Reactors that use graphite moderator include the Soviet-designed nuclear reactor that uses enriched uranium as its fuel, pebble bed reactors, and the Magnox reactor; a type of nuclear power/production reactor that was designed to run on natural uranium with graphite as the moderator and carbon dioxide gas as the heat exchange coolant.

Alternatives

Nuclear reactors can be either thermal or fast. Currently, almost all operating reactors are thermal and thus require a moderator to slow down fast neutrons to the thermal level so that nuclear fission can continue. However, in fast reactors a moderator is not needed, and the neutrons within it move much more quickly. Fast reactors are beneficial as they enhance the sustainability of nuclear power. This is because they have the ability to get more neutrons out of their fuel, can transform nuclear waste into products that decay more quickly, and they respond better to potentially catastrophic equipment failures. However, they are more expensive, and they can overheat fairly easily.

6.7.2 Moderators for Nuclear Rockets

It is clear that the above moderators will not work in the nuclear micro reactors planned for deep space propulsion. For example, the reactor of one of the contractors designs was only about the size of a large trash can; about 1m x 1.2m (3.28 ft x 4ft) and weighs approximately 3 tons. It uses beryllium oxide (BeO) as its moderator.

Background

L. N. Vauquelin first discovered beryllium (Be) while trying to identify the chemical composition of emerald and beryl in 1797. After that, researchers proposed various methods for the preparation of beryllium oxide (BeO. Although pure BeO was not available at that time, by 1922 its crystal structure and lattice parameters were determined. According to the published literature, the earliest mention of pure BeO was in research about the transmutation of elements by protons in 1933. Later in 1938, very pure BeO was obtained by firing beryllium sulfate (BeSO4) in a rotary kiln with a maximum temperature of 1450°C (2,642°F). Early studies provided evidence that the physical properties of BeO remained relatively stable over a wide range at elevated temperatures, which was quite different from other ceramic materials.

In 1943, H. C. Urey, a Manhattan project scientist, discussed the utilization of BeO as the tube material in a heavy water-slurry pile at Columbia University and it was the first time to put forward the utilization of BeO in nuclear reactors. However, this report mainly focused on the appropriate mechanical properties of BeO and the resistance to corrosion rather than the potential as the neutron moderator or reflector in nuclear reactors. In 1944, Dr. Farrington Daniels, another Manhattan project scientist, first proposed that BeO could replace graphite as a moderator when he put forward the new concept of high-temperature helium gas-cooled power reactors at the Metallurgical Laboratory of the University of Chicago.

Meanwhile, BeO functioning as the neutron reflecting material was first utilized in the famous water boiler reactor at Los Alamos National Laboratory in 1944. The preliminary design of a 12 MW Los Alamos reactor utilized approximately 12,250 kg (22,082°F) of BeO hexagonal blocks in 1945, in which BeO actually functioned as the moderator, the reflector and the diluent for the fuel system. However, this plan was interrupted since the AEC finally chose to support the development of light water reactors as large-scale civilian reactors in 1947.

From the 1950s–1980s, it was a golden age for the development of nuclear power worldwide. Many diverse concepts of nuclear reactors were put forward and tested experimentally. In the intervening years, BeO was extensively employed in these concepts

In the 1950s and 1960s, the AEC supported three major programs about nuclear propulsion: the nuclear propulsion project for manned bombers, nuclear rockets for space ships (See Chapter 2) and nuclear ramjets for unmanned bombers or missiles. Both the bomber and missile project utilized BeO as the neutron moderator or reflector. For instance, in 1950, the ORNL initiated the aircraft project in which BeO was reconsidered as the moderator and reflector material once more after the interruption of Dr. Daniels' reactor.

Over the past sixty years, there have been many studies and uses of BeO including the following:

- In 1958, the Maritime Gas-Cooled Reactor (MGCR) was initiated by the U. S. AEC and the Maritime Administration.
- Later, the Experimental Beryllium Oxide Reactor (EBOR) program became part of this MGCR program in 1961.
- In 1960, BeO was also selected as the diluent for UO_2 in the army gas-cooled reactor ML-1, which was a prototype of a mobile nuclear power reactor for use on ships or at remote locations. This UO_2single bond BeO system in the form of pellets stacked into fuel rods, possesses two major advantages: (1) better thermal conductivity due to the existence of BeO and (2) better capability of retaining the fission products.
- Nuclear reactors using the UO_2 single bond BeO pebble fuel system were reported in Australia, Unite Kingdom and Germany.
- In the 1960's, small compact nuclear reactors called Systems for Nuclear Auxiliary Power (SNAP), were designed for powering satellites and spacecraft for a long time; the first test flight of SNAP 10A was in 1965. BeO was employed in the SNAP as the control drum moderator material and the reflector.
- In the early 1970s, NASA developed a fast-spectrum thermionic reactor for generating electricity. A molybdenum-clad BeO reflector was employed in this reactor because of the limitation of the reactor size. In addition, BeO was also utilized in fast reactors and thermal breeder reactors as the moderator or reflector material in the 1970s.
- The French government also initiated a space nuclear powering program in 1982, in which 10 cm (3.94 in) thick BeO reflector slabs were placed above and below the reactor core.

In the past two decades of the 21st century, research on the utilization of BeO in nuclear reactors has attracted the attentions from various countries. Among these studies, BeO is most commonly employed in reactors as the neutron reflector .For instance; the Indian Kalpakkam Mini Reactor (KAMINI) utilizes U-233 as the fissile material, light water as the moderator and BeO as the reflector. Kilowatt Reactor Using Stirling Technology (KRUSTY) selected BeO as the reflector, which was developed by NASA and utilized a highly enriched uranium core and Stirling engines to generate electricity.

Finally, BeO as a solid-state moderator with prominent moderating performance is again valued in modern studies, since the active core moderated by BeO is normally physically smaller, which signifies that BeO still has the potential to be widely used in nuclear reactors, especially in micro nuclear reactors.

NASA's Moon to Mars exploration program will need nuclear propulsion and power systems for its Artemis program and crewed missions to Mars. See the book; *Artemis Base Camp: The First Steps.* Relying on solar energy alone is not enough, and a micro nuclear reactor is still the first and most viable option.

6.8 Reflectors

A neutron reflector is any material that reflects neutrons. While this sounds straight forward; it is not; in fact very complex papers have been written on the subject. To confuse things further, reflectors can also act as moderators. The subject is very germane to nuclear rocket reactors. To understand "reflection "; two terms need explanation; elastic and specular. In the simplest of terms, they are:

- Elastic scattering describes a process in which the total kinetic energy of the system is conserved. During elastic scattering of high-energy subatomic particles, linear energy transfer takes place until the incident particle's energy and speed has been reduced to the same as its surroundings, at which point the particle is "stopped".

- Specular reflection is a range of phenomena analogous to those observed in classical optics and are exhibited by slow neutrons; these include reflection, refraction and interference.

The reflector material may be graphite, beryllium, steel, tungsten carbide, gold, or other materials. A neutron reflector can make an otherwise subcritical mass of fissile material critical, or increase the amount of nuclear fission that a critical or supercritical mass will undergo. In nuclear reactors, the neutron's mean free path is critical as it undergoes elastic scattering on its way to becoming a slow-moving thermal neutron.

One way to think of a nuclear reflector is a layer of material immediately surrounding a reactor core that scatters back (or reflects) into the core many neutrons that would otherwise escape. The returned neutrons can then cause more fissions and improve the neutron economy of the reactor.

The Marshall Space Flight Center's Space-Capable Cryogenic Thermal Engine (SCCTE) project explored different ideas for the radial neutron reflector and control drums. The design under consideration is a radial reflector made of beryllium with dispersed cooling channels for flowing cryogenic hydrogen. Embedded in the reflector were sixteen control drums made from beryllium with boron carbide poison elements to control reactivity. The control drums have dispersed cooling channels for cryogenic hydrogen flow within the control drums. One potential design for the radial neutron reflector that meets the criteria of the SCCTE project

6.9 Heat Pipes

6.9.1 Fundamentals

Heat pipes are fundamental and necessary components of most spacecraft; Where ever there is heat; it will flow to the coldest point. In space, there is the absence of heat and the heat of the spacecraft components will want to flow to the cold of space. Consequently, there will always be the need to keep the spacecraft components warm; thus the need for heat pipes. They will definitely be included in the design of nuclear rockets and the spacecraft to the Moon and Mars.

While the subject of heat pipes seems straightforward, hundreds of research papers have been written over the past half century and scores of conferences have been held, attended by hundreds, if not thousands, of scientists. The discussions about various applications quickly become very complex as there is a lot of physics involved.

Heat pipes operate at a considerably higher temperature and pressure in microgravity than they do on Earth. The reason for this behavior is that there is no gravity-driven natural convection in microgravity, to help cool the heat pipe surface. The only way heat pipes can lose heat in space is to radiate that heat directly to the environment. It was found that the heat transfer coefficient in microgravity was higher comparable to the ground conditions.

The subject of heat pipe has attracted enormous interest among researchers and practitioners. New designs are invented for the cooling of space electronics and computers. Several millions of heat pipes of various configurations are being manufactured every month for cooling of laptop computers. Heat pipes are virtually everywhere; from the tiny ones in your laptop and phone to the huge ones along the 1,287 km (800 mi) long, 1.2 m (48 in) diameter Alaska pipeline. In between are the thousands of space applications.

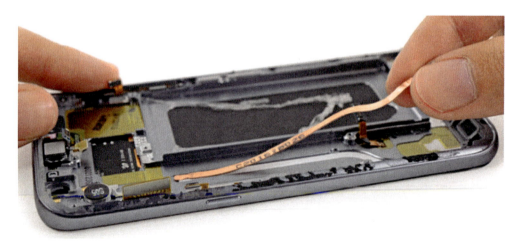

Fig. 6.8 A heat pipe application in a smart phone. Photo courtesy of Samsung.

Fig. 6.9 Heat pipes along the Alaska Pipeline. Photo courtesy of Luca Galuzzia

In this case, special heat pipes conduct heat from the oil to the fins at the top of the pipes in order to avoid thawing the permafrost.

6.9.2 Space Applications

The first application of heat pipes in the space program was the thermal equilibration of satellite transponders. As satellites orbit, one side is exposed to the direct radiation of the sun while the opposite side is completely dark and exposed to the deep cold of outer space. This causes severe temperature gradients affecting the reliability and accuracy of the transponders. The heat pipe cooling system designed for the purpose managed the high heat fluxes and demonstrated flawless operation with or without the gravitational influence. The cooling system developed was the first use of variable conductance heat pipes (VCHP) to actively regulate heat flow or evaporator temperature.

Heat pipes are very desirable components in the area of spacecraft cooling and temperature stabilization due to their low weight penalty, zero maintenance, and reliability. Maintaining isothermal structures is an important task with respect to, for example, orbiting astronomy experiments under the adverse solar heating. During orbit, an observatory is fixed on a single point such as a star. Therefore, one side of the spacecraft will be subjected to intense solar radiation, while the other is exposed to deep space. Heat pipes have been used to transport the heat from the side irradiated by the sun to the cold side in order to equalize the temperature of the structure. Heat pipes are also being used to dissipate heat generated by electronic components in satellites.

Early experiments of heat pipes for aerospace applications were conducted in sounding rockets which provided six to eight minutes of zero-g conditions. In 1974, ten separate heat pipe experiments were flown in the International Heat Pipe Experiment. Also heat pipe experiments were conducted aboard the Applications Technology Satellite-6, in which an ammonia heat pipe with a spiral artery wick was used as a thermal diode.

In the Space Shuttle era, flight testing of prototype heat pipe designs continued at a much larger scale. A 1.83 m (6 ft) mono groove heat pipe radiator with Freon 21 as the working fluid was flight tested on the eighth space shuttle flight. The International Space Station Heat Pipe Advanced Radiator Element (SHARE) consisting of a 15.2 m (50 ft) long high capacity mono groove heat pipe encased in a radiator panel, was flown on the Space Shuttle during 1989 and also, two heat pipe radiator panels were separately flight tested in a Space Shuttle flight of 1991.

Heat pipe thermal buses were proposed to facilitate a connection between heat-generating components and external radiator. The components may be designed with a clamping device which can be directly attached to the heat pipe thermal bus at various points in the spacecraft. In 1992, two different axially grooved oxygen heat pipes were tested in a Hitchhiker Canister experiment that was flown aboard the Shuttle Discovery (STS-53) by NASA and the Air Force to determine startup behavior and transport capabilities in micro gravity.

Heat pipes have also been qualified and/or used for thermal control applications in avionic systems including aircrafts with more electric architectures. There was a proposal to replace the radioisotope thermoelectric generating systems by the radioisotope Stirling systems as a long-lasting electricity generation solution in space missions due to their higher efficiency.

In the current radioisotope Sterling systems, if the Stirling engine stops, the heat removal from the system would be ceased and the insulation will be spoiled to prevent damage to the fuel, and the mission will be ended. The alkali-metal variable conductance heat pipes were proposed and tested to allow multiple stops and restarts of the Stirling engine

The Constrained Vapor Bubble experiment represents State-of-the-Art heat pipe research undertaken by NASA to cool the International Space Station (ISS). It is the prototype for a wickless heat pipe and is the first full scale fluids experiment flown on the U.S. in order to achieve a better understanding of the physics of evaporation and condensation and how they affect cooling processes in microgravity using a remotely controlled microscope and a small cooling device.

NASA Glenn Research Center has examined small fission power systems that address the gap between Radioisotope Power Systems (RPS) and Fission Surface Power Systems (FPS) for future spacecraft applications and Lunar and Martian surface missions. See Chapter 5. The Kilopower system, operating in the 1 to 10 kWe range, uses alkali metal heat pipes to supply heat to Stirling convertors to produce electricity and titanium water heat pipes to remove the waste heat and transport it to the radiators where it is rejected to space

6.9.3 Advanced Cooling Technologies, Inc. (ACT)

Continuing earlier work on heat pipes for fission power, Advanced Cooling Technologies developed a series of titanium-water heat pipe radiators to remove the waste heat from Kilopower system convertors. The titanium-water heat pipes have an advanced fluid management design, enabling the heat pipes and the Kilopower system to transport waste heat in space and surface operations and to survive and startup smoothly after being exposed to a frozen condition during launch or afterward.

The Kilopower system uses Stirling conversion to generate power. The thermal energy generated from a nuclear fission reactor is transferred to the Stirling convertor hot-end via a series of high temperature >800°C (1650°F)

alkali metal heat pipes. Parts of the thermal energy will be converted into usable electricity. The remaining waste heat needs to be removed from the Stirling convertor cold-end and ultimately rejected to the space environment through radiators.

ACT designed and fabricated multiple titanium-water heat pipes with radiator panels attached. Their thermal performance was validated through experimental measurement conducted both in ambient and a space-relevant environment (i.e. thermal vacuum chamber). The titanium-water heat pipes, operating at 400K (260°F) must be able to function and survive under the following four conditions:

1. Operating in space without gravity forces.

2. Operating on a planetary surface with a reduced gravity force for working fluid return.

3. Testing on the ground to estimate space operation performance. To do so, the heat pipe evaporator should be slightly higher (~ 0.1 inch) than the condenser.

4. Survival and recovery from a frozen condition. During a launch period, the heat pipes will be orientated in extreme against gravity and the sink temperature could be lower than the freezing point of the working fluid. It is necessary to incorporate a special wick design to manage the working fluid within the heat pipe, which can (a) avoid liquid staying inside the condenser and bursting the pipe while freezing and (b) supply enough amount of working fluid to start up the heat pipe after being frozen.

ACT developed a series of titanium water heat pipes for the Kilopower system cooling, based on our earlier titanium/water heat pipe work. The titanium water heat pipe has a C-shape evaporator to interface with the Stirling Convertor.

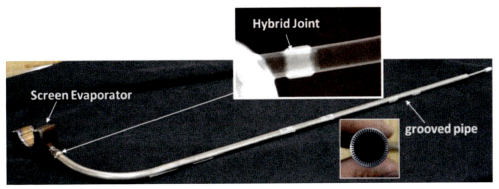

Fig. 6.10 Titanium Water Heat Pipe for Kilowatt. Photo courtesy of ACT, Inc.

Inside the evaporator, ACT inserted two types of screen mesh with different pore sizes which will enable the heat pipe to survive and recover from freezing. The rest of the pipe has an axial groove structure. Test results show that each titanium-water heat pipe is capable of transferring more than 400W of heat in the slightly adverse gravity inclination with a low thermal resistance at 0.01°C/W. The freeze/thaw test results further demonstrates that the heat pipe can successfully recover from a frozen condition at -50°C to a normal space operation mode.

ACT continues to support the NASA Glenn Research Center for support for many different space related projects.

For a 52:50 minute video by Advanced Cooling Technologies on the use of heat pipes in space applications, go to:

https://youtu.be/u4fMntL150A

IMAGE LINKS

Fig. 6.1 NEP Subsystems Graphic
https://www.researchgate.net/publication/349379840/figure/fig4/AS:99228214
4673797@1613589907936/Nuclear-electric-propulsion-subsystems-and-conceptual-design-SOURCE-Mars-Transportation.png

Fig. 6.2 Ion Thruster use on DART

https://www.syfy.com/sites/syfy/files/styles/blog-post-embedded--tablet-1_5x/public/2020/03/grc-2019-c-10067.jpg

Fig. 6.3 Hall Thruster NASA GRC
http://www.nasa.gov/sites/default/files/thumbnails/image/hall_thruster_0.jpg

Fig. 6.4 NTP Reactor Illustration
https://astronautical.org/dev/wp-content/uploads/2021/01/NPP-4.jpg

Fig. 6.5 Uranium Levels
https://www.centrusenergy.com/wp-content/uploads/2022/02/uranium-levels-1024x630.jpeg

Fig. 6.6 Centrus Demo Cascade
https://www.ans.org/file/9976/Picture1.jpg

Fig. 6.7 TRISO Nuclear Fuel diagram
https://art.inl.gov/trisofuels/SiteAssets/SitePages/Home/08-GA50711-01-R1.jpg

Fig. 6.8 Heat Pipe in a smart phone
https://9to5google.com/wp-content/uploads/sites/4/2016/03/galaxy-s7-ifixit-coil.jpg?quality=82&strip=all

Fig. 6.9 Heat Pipes along the Alaska Pipeline
https://lh3.googleusercontent.com/Y1ci5qnX3S0pCJFgl80dsPdUMBcQ75yyte SNCBsrLbYYnjZfKE5RKGmQbfqr1jzmNwETyX6138T_y4L63mvOBfKTa Z37irXjgDwAZAJvyQ

Fig. 6.10 Titanium Water Heat Pipe for Kilowatt
https://www.1-act.com/wp-content/uploads/2018/03/titaniumwaterheatpipe-768x278.jpg

7

Findings and Recommendations

Background

The National Academies of Science report on Space Nuclear Propulsion for Human Mars Exploration broke down their findings and recommendations into two categories; those related to nuclear thermal propulsion and nuclear electric propulsion. Keep in mind that this was published in 2021 and as of 2023, work as progressed in many of these areas. They represent many of the problems and concerns for acquiring the capabilities needed for deep space propulsion and the ability to provide the necessary power for space exploration.

7.1 NUCLEAR THERMAL PROPULSION

Finding: NTP Storage of LH2. NTP systems for the baseline mission will require long-duration storage of LH2 at 20 K with minimal boiloff in the vehicle assembly orbit and for the duration of the mission.

Recommendation: NTP Storage of LH2. If NASA plans to apply NTP technology to the baseline mission, NASA should develop high-capacity tank systems capable of storing LH2 at 20 K with minimal boiloff in the vehicle assembly orbit and for the duration of the mission.

Finding: Fuel Characterization. A significant amount of characterization of reactor core materials, including fuels, remains to be done before NASA and DOE will have sufficient information for a reactor core design.

Recommendation: NTP Fuel Architecture. If NASA plans to apply (NTP) technology to a 2039 launch of the baseline mission, NASA should expeditiously select and validate fuel architecture for an NTP system that is capable of achieving a propellant reactor exit temperature of approximately

$2700°$ K ($4,400°$F) or higher (which is the temperature that corresponds to the required Isp of 900 sec) without significant fuel deterioration during the mission lifetime. The selection process should consider whether the appropriate fuel feedstock production capabilities will be sufficient.

Finding: NTP Modeling and Simulation, Ground Testing, and Flight Testing. Subscale in-space flight testing of NTP systems cannot address many of the risks and potential failure modes associated with the baseline mission NTP system and therefore cannot replace full-scale ground testing. With sufficient M&S and ground testing of fully integrated systems, including tests at full scale and thrust, flight qualification requirements can be met by the cargo missions that will precede the first crewed mission to Mars.

Recommendation: NTP Modeling and Simulation, Ground Testing, and Flight Testing. To develop a nuclear thermal propulsion (NTP) system capable of executing the baseline mission, NASA should rely on (1) extensive investments in modeling and simulation; (2) ground testing, including integrated system tests at full scale and thrust; and (3) the use of cargo missions as a means of flight qualification of the NTP system that will be incorporated into the first crewed mission.

Finding: NTP Prospects for Program Success. An aggressive program could develop an NTP system capable of executing the baseline mission in 2039.

Recommendation: NTP Major Challenges. NASA should invigorate technology development associated with the fundamental nuclear thermal propulsion (NTP) challenge, which is to develop an NTP system that can heat its propellant to approximately $2700°$ K ($4,400°$F) at the reactor exit for the duration of each burn. NASA should also invigorate technology development associated with the long-term storage of liquid hydrogen in space with minimal loss, the lack of adequate ground-based test facilities, and the need to rapidly bring an NTP system to full operating temperature (preferably in 1 min or less).

7.2 NUCLEAR ELECTRIC PROPULSION

Finding: NEP Prospects for Program Success. As a result of low and intermittent investment over the past several decades, it is unclear if even an aggressive program would be able to develop an NEP system capable of executing the baseline mission in 2039.

Recommendation: NEP Major Challenges. NASA should invigorate technology development associated with the fundamental nuclear electric propulsion (NEP) challenge, which is to scale up the operating power of each NEP subsystem and to develop an integrated NEP system suitable for the baseline mission. In addition, NASA should put in place plans for (1) demonstrating the operational reliability of an integrated NEP system over its multiyear lifetime and (2) developing a large-scale chemical propulsion system that is compatible with NEP.

Recommendation: NEP Pace of Technology Development. If NASA plans to apply nuclear electric propulsion (NEP) technology to a 2039 launch of the baseline mission, NASA should immediately accelerate NEP technology development.

7.3 FINDINGS AND RECOMMENDATIONS APPLICABLE TO BOTH NTP AND NEP SYSTEMS

Finding: Trade Studies. Recent, apples-to-apples trade studies comparing NEP and NTP systems for a crewed mission to Mars in general and the baseline mission in particular, do not exist.

Recommendation: Trade Studies. NASA should develop consistent figures of merit and technical expertise to allow for an objective comparison of the ability of nuclear electric propulsion and nuclear thermal propulsion systems to meet requirements for a 2039 launch of the baseline mission.

Finding: NEP and NTP Commonalities. NEP and NTP systems require, albeit to different levels, significant maturation in areas such as nuclear reactor fuels, materials, and additional reactor technologies; cryogenic fluid management; M&S; testing; safety; and regulatory approvals. Given these commonalities, some development work in these areas can proceed independently of the selection of a particular space nuclear propulsion system.

Finding: Enrichment of Nuclear Fuels. A comprehensive assessment of HALEU versus HEU for NTP and NEP systems that weighs the key considerations is not available. These considerations include technical feasibility and difficulty, performance, proliferation and security, safety, fuel availability, cost, schedule, and supply chain as applied to the baseline mission. **Recommendation:** Enrichment of Nuclear Fuels. In the near term, NASA and the Department of Energy, with inputs from other key stakeholders, including commercial industry and academia, should conduct a comprehensive assessment of the relative merits and challenges of highly enriched uranium and high-assay, low-enriched uranium fuels for nuclear thermal propulsion and nuclear electric propulsion systems as applied to the baseline mission.

Finding: Synergies with Terrestrial and National Defense Nuclear Systems. Terrestrial micro reactors, which operate at a power level comparable to NEP reactors, are on a faster development and demonstration timeline than current plans for space nuclear propulsion systems. Development of micro reactors may provide technology advances and lessons learned relevant to the development of NEP systems. Similarly, technology advances within the DARPA DRACO program could potentially contribute to the development of NTP systems for the baseline mission. **Recommendation:** Synergies with Terrestrial and National Defense Nuclear Systems. NASA should seek opportunities for collaboration with the Department of Energy and Department of Defense terrestrial micro-reactor programs and the Defense Advanced Research Projects Agency DRACO program to identify synergies with NASA space nuclear propulsion programs.

8

International NTP/NEP Efforts

8.1 ESA

ESA's concept of a space logistics ecosystem is a system-of-systems supported by a tailored portfolio of interconnected In-Space Transportation Vehicles (ISTVs) based on standardized interfaces.

ESA notes that the development of new and efficient space propulsion systems is necessary as propulsion is at the heart of the space logistics. Without propulsion, no logistics and no services could be carried out in orbit (or in deep space). In addition, efficient propulsion systems minimize the environmental impacts; less launch weight to space, less propellant needed maximizes the performance.

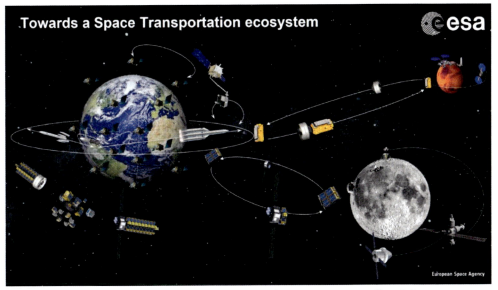

Fig. 8.1 The concept of a Space Transportation Ecosystem. Graphic courtesy of ESA.

Today, in-space propulsion is either chemical with storable propellant or electric with solar power. In the future, chemical propulsion will use cryogenic propellant and electric propulsion will handle large power inputs. However, these two technologies are getting close to their physical limits, beyond which performance increase is impossible. To pass by these limits and enable the space logistics to a new league, a new type of propulsion shall be introduced: nuclear based propulsion systems.

While most of previous studies in Europe on nuclear propulsion were terminated without sequel, today's technologies are making nuclear propulsion a plausible alternative to conventional propulsion systems. Nuclear propulsion can be multiple times more efficient than the most efficient chemical propulsion, or exceed the electric power limited by solar energy, thus enabling exploration where no other technology can go.

The ESA Future Space Transportation Systems department (STS-F) recently initiated two feasibility studies. These studies identified key technologies to mature and intermediate steps to take to develop nuclear propulsion in Europe. Of course, the essential conditions to use nuclear energy for propulsion are the robust implementation of the safety requirements from the early phases of the design. Following recommendations established in 2009 from the International Atomic Energy Agency (IAEA) and United Nation, the safety aspect is at the center of the development of nuclear technologies for space applications.

The most important reason for NEP is to explore beyond Mars orbit where solar power is limited. To develop a "tug" for long term and distant exploration, a stepwise implementation could be considered with a subscale tug to in-orbit demonstrating the added value of NEP with less demanding missions.

ESA is studying a first step toward an in-orbit demonstrator of nuclear electric propulsion systems. The main advantage over chemical reaction is the efficiency of the engines. The advantage over solar electric power input is the larger power output and independence of exposure to direct sunlight, especially for transporting heavy cargo with long time constraints and for exploration beyond Mars orbit.

In addition, NEP could have strong synergies with other space application. For instance, nuclear power could be used on the Moon or Mars surface to power future habitats or robotic exploration of the solar system, or in space for other purpose than propulsion. On the propulsion side, the development of high-power electric thrusters for NEP could also be used with non-nuclear power input.

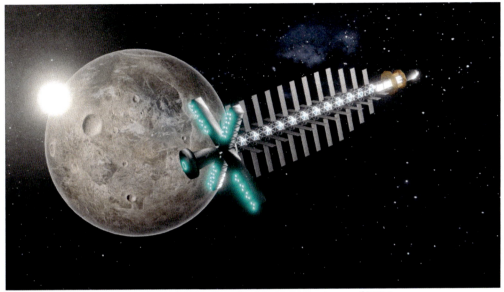

Fig. 8.2 ESA concept for a NEP "tug". Photo courtesy of ESA.

The European Space Agency (ESA) has awarded a contract to Belgian company Tractebel to evaluate the possibility of producing plutonium-238 (Pu-238) for use in space exploration. Radioisotope power sources, sometimes referred to as nuclear batteries fueled with Pu-238 have been used in space missions since the early 1960s. Tractebel will study the possibility of manufacturing Pu-238 by bombarding neptunium-237 from the La Hague recycling facility in France with the neutron flux of the BR2 research reactor in Mol, Belgium. The work will include the development of a "roadmap" for the creation of a Pu-238 production chain in Europe that will include a timeline and estimated production capacity and costs, as well as evaluating regulatory acceptance.

ESA is manufacturing the European Large Logistic Lander (EL3) to land on the Moon as part of Europe's participation in the NASA-led Lunar Gateway program.

8.2 UNITED KINGDOM (UK)

On March 17, 2023, the UK Space Agency (UKSA) announced £2.9 million of new funding for the project which will deliver an initial demonstration of a UK lunar modular nuclear reactor to Rolls-Royce. This follows a £249,000 study funded by the UK Space Agency in 2022.

Scientists and engineers at Rolls-Royce are working on the Micro-Reactor programme to develop technology that will provide power needed for humans to live and work on the Moon. All space missions depend on a power source, to support systems for communications, life-support and science experiments. Nuclear power has the potential to dramatically increase the duration of future Lunar missions and their scientific value.

The UK Space Agency's Minister of State at the Department of Science, Innovation and Technology, George Freeman, said: "Space exploration is the ultimate laboratory for so many of the transformational technologies we need on Earth: from materials to robotics, nutrition, clean-tech and much more. As we prepare to see humans return to the Moon for the first time in more than 50 years, we are backing exciting research like this lunar modular reactor with Rolls-Royce to pioneer new power sources for a lunar base. Partnerships like this, between British industry, the UK Space Agency and government are helping to create jobs across our £16 billion Space Tech sector and help ensure the UK continues to be a major force in frontier science".

Fig. 8.3 Rolls-Royce Micro-Reactor on the Moon. Photo courtesy of Rolls-Royce.

Nuclear space power is anticipated to create new skilled jobs across the UK to support the burgeoning UK space economy. Rolls-Royce plan to have a reactor ready to send to the Moon by 2029. Relatively small and lightweight compared to other power systems, a nuclear micro-reactor could enable continuous power regardless of location, available sunlight, and other environmental conditions.

Rolls-Royce will be working alongside a variety of collaborators including the University of Oxford, University of Bangor, University of Brighton, University of Sheffield's Advanced Manufacturing Research Centre (AMRC) and Nuclear AMRC. The funding means Rolls-Royce can further strengthen its knowledge of these complex systems, with a focus on three key features of the Micro-Reactor; the fuel used to generate heat, the method of heat transfer and technology to convert that heat into electricity.

The potential applications of Rolls-Royce Micro-Reactor technology are wide-ranging and could support commercial and defence use cases in addition to those in space. The aim is to create a world-leading power and propulsion capability for multiple markets and operator needs, alongside a clean, green and long-term power source.

Abi Clayton, Director of Future Programmes for Rolls-Royce said: "The new tranche of funding from the UK Space Agency means so much for the Rolls-Royce Micro-Reactor Programme. We're proud to work collaboratively with the UK Space Agency and the many UK academic institutions to showcase the best of UK innovation and knowledge in space. This funding will bring us further down the road in making the Micro-Reactor a reality, with the technology bringing immense benefits for both space and Earth. The technology will deliver the capability to support commercial and defence use cases alongside providing a solution to decarbonise industry and provide clean, safe and reliable energy".

Dr Paul Bate, Chief Executive of the UK Space Agency, said: "We are backing technology and capabilities to support ambitious space exploration missions and boost sector growth across the UK. Developing space nuclear power offers a unique chance to support innovative technologies and grow our nuclear, science and space engineering skills base".

This innovative research by Rolls-Royce could lay the groundwork for powering continuous human presence on the Moon, while enhancing the wider UK space sector, creating jobs and generating further investment. The partnership with Rolls-Royce comes after the UK Space Agency recently announced £51 million of funding available for UK companies to develop communication and navigation services for missions to the Moon, as part of the European Space Agency's Moonlight programme, which aims to launch a constellation of satellites into orbit around the Moon. This will allow future astronauts, rovers, science experiments and other equipment to communicate, share large amounts of data including high-definition video, and navigate safely across the lunar surface.

8.3 Other Countries

Roscosmos is also pursuing NEP technology with its Transport and Energy Module (TEM) program; a Russian project to create an uncrewed nuclear electric rocket spaceship for Solar System exploration. The first orbital flight test of the reactor is planned for no earlier than 2030.The first mission, named Zeus, is envisioned to operate for 50 months and deliver payloads to the Moon, Venus, and Jupiter through multiple gravity assists. See Appendix 6.

In 2017, the China Aerospace Science and Technology Corporation (CASC); the Chinese National Space Agency's (CNSA) main contractor, released its "Space Transportation Roadmap." In addition to the development of a reusable spaceplane (similar to the X-37B), this document also called for the creation of a single-stage-to-orbit (SSTO) spaceplane and fully-reusable rockets by the 2030s, and a nuclear-powered shuttle by 2045. See Appendix 7.

In 2022, it was reported that a Chinese nuclear reactor for providing power and propulsion in outer space passed a comprehensive performance evaluation. The reactor was designed by the Chinese Academy of Sciences and can generate one megawatt of electricity for spacecraft power supply and propulsion. The project was initiated in 2019 as a national key research and development program and demonstrates a strong Chinese interest in developing nuclear power for use in space.

China has been expanding its space transportation and deep space capabilities in recent years, successfully developing cryogenic rockets to facilitate lunar, Mars and space station projects. It is now working on reusable launchers, super heavy-lift rockets and a two-stage-to-orbit reusable spaceplane system.

The Indian Space Research Organisation (ISRO) is exploring the possibility of developing a new propulsion technology for its future deep space missions. "design and modelling; simulation and analysis; testing and "design and modelling; simulation and analysis; testing and qualification of 100W Radioisotope Thermoelectric Generator.

Between NASA, the ESA, Roscosmos, China, and India, there are some very ambitious plans for space exploration in the coming decades. These include returning to the Moon and building bases that would allow for a sustained human presence. Between the 2030s and 2040s, all the major space players hope to have placed footprints (and possibly habitats) on Mars as well.

IMAGE LINKS

Fig. 8.1 The concept of a Space Transportation Ecosystem. Graphic courtesy of ESA.
https://commercialisation.esa.int/wp-content/uploads/2022/11/space-transport.jpg

Fig. 8.2 ESA concept for a NEP "tug". Photo courtesy of ESA.
https://commercialisation.esa.int/wp-content/uploads/2022/09/NEP-ROCKETROLL-1.jpg

Fig. 8.3 Rolls Royce Micro-Reactor
https://assets.publishing.service.gov.uk/government/uploads/system/uploads/image_data/file/178830/resized_1.png

9

Conclusions

In the Socratic dialogue, "The Republic", Plato famously wrote: "our need will be the real creator" which was molded over time into the English proverb; "Necessity is the mother of invention". I have interrupted this to mean; the need, is to get to Mars more quickly, and the invention is the Nuclear Rocket.

We have seen that NASA and the supporting government agencies have spent billions over decades developing nuclear rocket technologies and capabilities. But the time was just not conducive to completion of the task. The Vietnam War, the Apollo Program and societal demands were taking up all the resources; the vision of going to Mars was too faint. Consequently, no complete nuclear thermal propulsion (NTP) system has ever been assembled and tested in its flight configuration or flown in space. Other NTP programs have been carried out since Rover/NERVA, but only a few have built any prototype reactors or engines.

Fortunately, time has been good for the advancement of technology in many areas. We can almost laugh at the analog world of yesterday. Today, we can't image a world without the Internet; without a digital world that gives us immediate access to information. In hindsight, we definitely did not have the capability to go to Mars in those days; we barely have the capability today. In our mind's eye, we can be there in an instant.

But, it is clear that the lessons learned from the Rover and NERVA programs of the past are still viable and valuable. Those laboratory and test facilities are still here and are even more advanced today. They have the data from yesteryear. They have not been idle. They have continued with the advancement of nuclear thermal propulsion, nuclear electric propulsion, nuclear fuel and nuclear surface power technologies as well as have produced prototypes of flight hardware.

In 2011, there was a renewed interest and resurgence in deep space exploration. It was obvious to many that, while chemical propellants had their place in getting megatons out of the deep gravity well of the Earth, it just wouldn't do for the rest of the trip to Mars. It was alright for the nearby Moon, as was proven before, but not for really deep space missions. It was estimated that using nuclear thermal propulsion would save up to four Space Launch System launches for a human mission to Mars and save $B's, as well as to shorten the total launch schedule, the total mission duration, and to decrease the physical loads on the crew and thereby, increase the chances of mission success.

The National Academies of Science Report on Space Nuclear Propulsion for Human Mars Exploration examined the Baseline Missions to Mars in 2039, and broke down their findings and recommendations into two categories; those related to nuclear thermal propulsion and those related to nuclear electric propulsion. In 2021, they presented their findings and recommendations to NASA. They address many of the problems and concerns for acquiring the capabilities needed for deep space propulsion and the ability to provide the necessary power for space exploration.

The Report recommended NASA should:

- Develop high-capacity tank systems capable of storing LH2 at 20° K (-424°F) with minimal boiloff in the vehicle assembly orbit and for the duration of the mission.
- Expeditiously select and validate fuel architecture for an NTP system that is capable of achieving a propellant reactor exit temperature of approximately 2700° K (4,400°F) or higher without significant fuel deterioration during the mission lifetime. The selection process should consider whether the appropriate fuel feedstock production capabilities will be sufficient.
- Rely on (1) extensive investments in modeling and simulation; (2) ground testing, including integrated system tests at full scale and thrust; and (3) the use of cargo missions as a means of flight qualification of the NTP system that will be incorporated into the first crewed mission.

- Invigorate technology development associated with the fundamental nuclear electric propulsion (NEP) challenge, which is to scale up the operating power of each NEP subsystem and to develop an integrated NEP system suitable for the baseline mission.
- Put in place plans for demonstrating the operational reliability of an integrated NEP system over its multiyear lifetime and developing a large-scale chemical propulsion system that is compatible with NEP.
- Develop consistent figures of merit and technical expertise to allow for an objective comparison of the ability of nuclear electric propulsion and nuclear thermal propulsion systems to meet requirements for a 2039 launch of the baseline mission.
- Conduct a comprehensive assessment of the relative merits and challenges of highly enriched uranium and high-assay, low-enriched uranium fuels for nuclear thermal propulsion and nuclear electric propulsion systems as applied to the baseline mission.
- Should seek opportunities for collaboration with the DOE and DOD terrestrial micro-reactor programs and the DARPA DRACO program to identify synergies with NASA space nuclear propulsion programs.

Major efforts are now underway within NASA, the DOD and DOE as well as with their selected contractors, to meet the challenges laid out by the above recommendations. These efforts involve the NASA Centers, the DOD's DARPA, and the DOE's National Laboratories and the supporting contractors. Contracts have been let to three industry teams for both the nuclear thermal propulsion and nuclear fission surface power areas. These efforts cover many technological areas that have advanced the state-of-the-art or advanced their levels of flight readiness for the systems needed for a Mars mission. In many cases, commercial applications have already been derived from this research and testing. For example, micro reactors, using the same special fuels for space, have already been applied to the commercial section for specialized use in communities and industry.

To be sure, the space capable nations are also working on the problems of deep space exploration. While most countries are not capable of flights to the Moon; let alone to Mars, they are team players in supplying contributing elements. Even our Artemis Program is supported by the 22 member nations of ESA, the UK and others. The first flight to Mars will have many contributing countries with their best contractors wanting to contribute. For example, even traditional, well- known companies like Toyota and Mercedes Benz are working on space exploration. Space exploration needs to go deep into society's industrial base for support in order to be successful.

At the heart, or should I say the core, of nuclear rockets for deep space exploration is a very unique element; uranium. It is the last of the naturally occurring elements; #92 on the periodic table. It is just hard to believe that so little uranium can do so much. According to a technical report drafted by Dr. Michael G. Houts, the NTP Principal Investigator at NASA Marshall, an NTP rocket could generate 200 kWt of power using a single kilogram of uranium for a period of 13 years. That's twice the efficiency of chemical rockets. At that rate, a nuclear thermal rocket could make the trip to Mars in half the time! Another nuclear scientist put it this way; the envisioned rocket engine that NASA and its team is developing to go to Mars, would only require a volume of uranium that is roughly the physical size of a marble! Still another company states that their micro reactor design is only about the size of a large garbage can; about 1m x 1.2 m (3.28 ft x 4ft).

Can it be that mankind has finally harnessed the power of the atom to the point where humans can actually go to Mars?

Appendix 1

Space Policy Directive (SPD) 6, titled "National Strategy for Space Nuclear Power and Propulsion,"

Memorandum on the National Strategy for Space Nuclear Power and Propulsion (Space Policy Directive-6)

INFRASTRUCTURE & TECHNOLOGY

Issued on: December 16, 2020
Memorandum for the Vice President
The Secretary of State
The Secretary of Defense
The Secretary of Commerce
The Secretary of Transportation
The Secretary of Energy
The Director of the Office of Management and Budget
The Assistant to the President for National Security Affairs
The Administrator of the National Aeronautics and Space Administration
The Chairman of the Nuclear Regulatory Commission
The Director of the Office of Science and Technology Policy

SUBJECT: National Strategy for Space Nuclear Power and Propulsion

Section 1. Policy. The ability to use space nuclear power and propulsion (SNPP) systems safely, securely, and sustainably is vital to maintaining and advancing United States dominance and strategic leadership in space. SNPP systems include radioisotope power systems (RPSs) and fission reactors used for power or propulsion in spacecraft, rovers, and other surface elements. SNPP systems can allow operation of such elements in environments in which solar and chemical power are inadequate. They can produce more power at lower mass and volume compared to other energy sources, thereby enabling persistent presence and operations. SNPP systems also can shorten transit times for crewed and robotic spacecraft, thereby reducing radiation exposure in harsh space environments.

National Security Presidential Memorandum-20 (NSPM-20) of August 20, 2019 (Launch of Spacecraft Containing Space Nuclear Systems) updated the process for launches of spacecraft containing space nuclear systems. It established it as the policy of the United States to "develop and use space nuclear systems when such systems safely enable or enhance space exploration or operational capabilities."

Cooperation with commercial and international partners is critical to achieving America's objectives for space exploration. Presidential Policy Directive 4 of June 28, 2010 (National Space Policy), as amended by the Presidential Memorandum of December 11, 2017 (Reinvigorating America's Human Space Exploration Program), established it as the policy of the United States to "[l]lead an innovative and sustainable program of exploration with commercial and international partners to enable human expansion across the solar system and to bring back to Earth new knowledge and opportunities."

This memorandum establishes a national strategy to ensure the development and use of SNPP systems when appropriate to enable and achieve the scientific, exploration, national security, and commercial objectives of the United States. In the context of this strategy only, the term "development" includes the full development process from design through testing and production, and the term "use" includes launch, operation, and disposition. This memorandum outlines high-level policy goals and a supporting roadmap that will advance the ability of the United States to use SNPP systems safely, securely, and sustainably. The execution of this strategy will be subject to relevant budgetary and regulatory processes and to the availability of appropriations.

Sec. 2. Goals. The United States will pursue goals for SNPP development and use that are both mission-enabling and ambitious in their substance and their timeline. These goals will enable a range of existing and future space missions, with the aim of accelerating achievement of key milestones, including in-space demonstration and use of new SNPP capabilities. This memorandum establishes the following such goals for the Nation:

(a) Develop uranium fuel processing capabilities that enable production of fuel that is suitable to lunar and planetary surface and in-space power, nuclear electric propulsion (NEP), and nuclear thermal propulsion (NTP) applications, as needed. These capabilities should support the ability to produce different uranium fuel forms to meet the nearest-term mission needs and, to the extent feasible, should maximize commonality — meaning use of the same or similar materials, processes, designs, or infrastructure — across these fuel forms. To maximize private-sector engagement and cost savings, these capabilities should be developed to enable a range of terrestrial as well as space applications, including future commercial applications;

(b) Demonstrate a fission power system on the surface of the Moon that is scalable to a power range of 40 kilowatt-electric (kWe) and higher to support a sustained lunar presence and exploration of Mars. To the extent feasible, this power system should align with mission needs for, and potential future government and commercial applications of, in-space power, NEP, and terrestrial nuclear power;

(c) Establish the technical foundations and capabilities — including through identification and resolution of the key technical challenges — that will enable options for NTP to meet future Department of Defense (DoD) and National Aeronautics and Space Administration (NASA) mission requirements; and

(d) Develop advanced RPS capabilities that provide higher fuel efficiency, higher specific energy, and longer operational lifetime than existing RPS capabilities, thus enabling survivable surface elements to support robotic and human exploration of the Moon and Mars and extending robotic exploration of the solar system.

Sec. 3. Principles. The United States will adhere to principles of safety, security, and sustainability in its development and use of SNPP systems, in accordance with all applicable Federal laws and consistent with international obligations and commitments.

(a) Safety. All executive departments and agencies (agencies) involved in the development and use of SNPP systems shall take appropriate measures to ensure, within their respective roles and responsibilities, the safe development, testing, launch, operation, and disposition of SNPP systems. For United States Government SNPP programs, the sponsoring agency holds primary responsibility for safety. For programs involving multiple agencies, the terms of cooperation shall designate a lead agency with primary responsibility for safety in each stage of development and use.

(i) Ground development. Activities associated with ground development, including ground testing, of SNPP systems shall be conducted in accordance with applicable Federal, State, and local laws and existing authorities of regulatory agencies.

(ii) Launch. NSPM-20 established safety guidelines and safety analysis and review processes for Federal Government launches of spacecraft containing space nuclear systems, including SNPP systems, and for launches for which the Department of Transportation has statutory authority to license as commercial space launch activities (commercial launches). These guidelines and processes address launch and any subsequent stages during which accidents may result in radiological effects on the public or the environment — for instance, in an unplanned reentry from Earth orbit or during an Earth flyby. Launch activities shall be conducted in accordance with these guidelines and processes.

(iii) Operation and disposition. The operation and disposition of SNPP systems shall be planned and conducted in a manner that protects human and environmental safety and national security assets. Fission reactor SNPP systems may be operated on interplanetary missions, in sufficiently high orbits, and in low-Earth orbits if they are stored in sufficiently high orbits after the operational part of their mission. In this context, a sufficiently high orbit is one in which the orbital lifetime of the spacecraft is long enough for the fission products to decay to a level of radioactivity comparable to that of uranium-235 by the time it reenters the Earth's atmosphere, and the risks to existing and future space missions and of collision with objects in space are minimized. Spacecraft operating fission reactors in low-Earth orbits shall incorporate a highly reliable operational system to ensure effective and controlled disposition of the reactor.

(b) Security. All agencies involved in the development and use of SNPP systems shall take appropriate measures to protect nuclear and radiological materials and sensitive information, consistent with sound nuclear nonproliferation principles. For United States Government SNPP programs, the sponsoring agency holds primary responsibility for security. For programs involving multiple agencies, the terms of cooperation shall designate a lead agency with primary responsibility for security in each stage of development and use. The use of highly enriched uranium (HEU) in SNPP systems should be limited to applications for which the mission would not be viable with other nuclear fuels or non-nuclear power sources. Before selecting HEU or, for fission reactor systems, any nuclear fuel other than low-enriched uranium (LEU), for any given SNPP design or mission, the sponsoring agency shall conduct a thorough technical review to assess the viability of alternative nuclear fuels. The sponsoring agency shall provide to the respective staffs of the National Security Council, the National Space Council, the Office of Science and Technology Policy, and the Office of Management and Budget a briefing that provides justification for why the use of HEU or other non-LEU fuel is required, and any steps the agency has taken to address nuclear safety, security, and proliferation-related risks. The Director of the Office of Science and Technology Policy shall ensure, through the National Science and

Technology Council, that other relevant agencies are invited to participate in these briefings.

(c) Sustainability. All agencies involved in the development and use of SNPP systems shall take appropriate measures to conduct these activities in a manner that is suitable for the long-term sustainment of United States space capabilities and leadership in SNPP.

(i) Coordination and Collaboration. To maximize efficiency and return on taxpayer investment, the heads of relevant agencies shall seek and pursue opportunities to coordinate among existing and future SNPP development and use programs. Connecting current efforts with likely future applications will help ensure that such programs can contribute to long-term United States SNPP capabilities and leadership. Agencies also shall seek opportunities to partner with the private sector, including academic institutions, in order to facilitate contributions to United States SNPP capabilities and leadership. To help identify opportunities for collaboration, the heads of relevant agencies should conduct regular technical exchanges among SNPP programs, to the extent that such exchanges are consistent with the principle of security and comply with applicable Federal, State, and local laws. Agencies shall coordinate with the Department of State when seeking opportunities for international partnerships.

(ii) Commonality. The heads of relevant agencies shall seek to identify and use opportunities for commonality among SNPP systems, and between SNPP and terrestrial nuclear systems, whenever doing so could advance program and policy objectives without unduly inhibiting innovation or market development, or hampering system suitability to specific mission applications. For example, opportunities for commonality may exist in goals (e.g., demonstration timeline), reactor design, nuclear fuels (e.g., fuel type and form, and enrichment level), supplementary systems (e.g., power conversion, moderator, reflector, shielding, and system vessel), methods (e.g., additive manufacturing of fuel or reactor elements), and infrastructure (e.g., fuel supply, testing facilities, launch facilities, and workforce).

(iii) Cost-effectiveness. The heads of relevant agencies should pursue SNPP development and use solutions that are cost-effective while also consistent with the principles of safety and security. For any program or system, the heads of such agencies should seek to identify the combination of in-space and ground-based testing and certification that will best qualify the system for a given mission while ensuring public safety.

Sec. 4. Roles and Responsibilities. (a) The Vice President, on behalf of the President and acting through the National Space Council, shall coordinate United States policy related to use of SNPP systems.

(b) The Secretary of State shall, under the direction of the President, coordinate United States activities related to international obligations and commitments and international cooperation involving SNPP.

(c) The Secretary of Defense shall conduct and support activities associated with development and use of SNPP systems to enable and achieve United States national security objectives. When appropriate, the Secretary of Defense shall facilitate private-sector engagement in DoD SNPP activities.

(d) The Secretary of Commerce shall promote responsible United States commercial SNPP investment, innovation, and use, and shall, when consistent with the authorities of the Secretary, ensure the publication of clear, flexible, performance-based rules that are applicable to use of SNPP and are easily navigated. Under the direction of the Secretary of Commerce, the Department of Commerce (DOC) shall ascertain and communicate the views of private-sector partners and potential private-sector partners to relevant agency partners in order to facilitate public-private collaboration in SNPP development and use.

(e) The Secretary of Transportation's statutory authority includes licensing commercial launches and reentries, including vehicles containing SNPP systems. Within this capacity, the Secretary of Transportation shall, when appropriate, facilitate private-sector engagement in the launch or reentry aspect of SNPP development and use activities, in support of United States science, exploration, national security, and commercial objectives. To help ensure the launch safety of an SNPP payload, and consistent with 51 U.S.C. 50904, a payload review may be conducted as part of a license application review or may be requested by a payload owner or operator in advance of or apart from a license application.

(f) The Secretary of Energy shall, in coordination with sponsoring agencies and other agencies, as appropriate, support development and use of SNPP systems to enable and achieve United States scientific, exploration, and national security objectives. When appropriate, the Secretary of Energy shall work with sponsoring agencies and DOC to facilitate United States private-sector engagement in Department of Energy (DOE) SNPP activities. Under the direction of the Secretary of Energy and consistent with the authorities granted to DOE, including authorities under the Atomic Energy Act of 1954 (AEA), as amended, 42 U.S.C. 2011, et seq., DOE may authorize ground-based SNPP development activities, including DOE activities conducted in coordination with sponsoring agencies and private-sector entities. As directed in NSPM-20, the Secretary of Energy shall maintain, on a full-cost recovery basis, the capability and infrastructure to develop, furnish, and conduct safety analyses for space nuclear systems for use in United States Government space systems.

(g) The Administrator of NASA shall conduct and support activities associated with development and use of SNPP systems to enable and achieve United States space science and exploration objectives. The Administrator of NASA shall establish the performance requirements for SNPP capabilities necessary to achieve those objectives. When appropriate, the Administrator of NASA shall facilitate private-sector engagement in NASA SNPP activities, and shall coordinate with the Secretary of Commerce and, as appropriate, the Secretary of State and the Secretary of Energy, to help facilitate private-sector SNPP activities.

(h) The Nuclear Regulatory Commission (NRC) has statutory authority under the AEA for licensing and regulatory safety and security oversight of commercial nuclear activities taking place within the United States. The NRC should, as appropriate and particularly in circumstances within NRC authority where DOE regulatory authorities cannot be applied, enable private-sector engagement in SNPP development and use activities in support of United States science, exploration, national security, and commercial objectives.

(i) The Director of the Office and Science and Technology Policy shall coordinate United States policy related to research and development of SNPP systems.

Sec. 5. Roadmap. The United States will pursue a coordinated roadmap for federally-supported SNPP activities to achieve the goals and uphold the principles established in this memorandum. This roadmap comprises the following elements, which the relevant agencies should pursue consistent with the following objective timeline, subject to relevant budgetary and regulatory processes and to the availability of appropriations:

(a) By the mid-2020s, develop uranium fuel processing capabilities that enable production of fuel that is suitable for lunar and planetary surface and in-space power, NEP, and NTP applications, as needed.

(i) Identify relevant mission needs. DoD and NASA should provide to DOE any mission needs (e.g., power density, environment, and timelines) relevant to the identification of fuels suitable for planetary surface and in-space power, NEP, and NTP applications.

(ii) Identify candidate fuel or fuels. DoD and NASA, in cooperation with DOE and private-sector partners, as appropriate, should identify candidate fuel or fuels to meet the identified mission requirements. This review and assessment should account for current and expected United States capabilities to produce and qualify for use candidate fuels, and for potential commonality of fuels or fuel variants across multiple planetary surface and in-space power, in-space propulsion, and terrestrial applications.

(iii) Qualify at least one candidate fuel. DoD and NASA, in cooperation with DOE and private-sector partners, as appropriate, should qualify a fuel or fuels for demonstrations of a planetary surface power reactor and an in-space propulsion system. While seeking opportunities to use private-sector-partner capabilities, agencies should ensure that the Federal Government retains an ability for screening and qualification of candidate fuels.

(iv) Supply fuel for demonstrations. DOE, in cooperation with NASA and DoD, and with private-sector partners, as appropriate, should identify feedstock and uranium that can be made available for planetary surface power and in-space propulsion demonstrations. DOE shall ensure that any provision of nuclear material for SNPP will not disrupt enriched uranium supplies for the United States nuclear weapons program and the naval propulsion program, and that SNPP needs are included among broader considerations of nuclear fuel supply provisioning and management.

(b) By the mid- to late-2020s, demonstrate a fission power system on the surface of the Moon that is scalable to a power range of 40 kWe and higher to support sustained lunar presence and exploration of Mars.

(i) Initiate a surface power project. NASA should initiate a fission surface power project for lunar surface demonstration by 2027, with scalability to Mars exploration. NASA should consult with DoD and other agencies, and with the private sector, as appropriate, when developing project requirements.

(ii) Conduct technology and requirements assessment. NASA, in coordination with DoD and other agencies, and with private-sector partners, as appropriate, should evaluate technology options for a surface power system including reactor designs, power conversion, shielding, and thermal management. NASA should work with other agencies, and private-sector partners, as appropriate, to evaluate opportunities for commonality among other SNPP needs, including in-space power and terrestrial power needs, possible NEP technology needs, and reactor demonstrations planned by NASA, other agencies, or the private sector.

(iii) Engage the private sector. DOE and NASA should determine a mechanism or mechanisms for engaging with the private sector to meet NASA's SNPP surface power needs in an effective manner consistent with the guiding principles set forth in this memorandum. In evaluating mechanisms, DOE and NASA should consider the possibility of NASA issuing a request for proposal for the development and construction of the surface power reactor system or demonstration.

(iv) System development. NASA should work with DOE, and with other agencies and private-sector partners, as appropriate, to develop the lunar surface power demonstration project.

(v) Conduct demonstration mission. NASA, in coordination with other agencies and with private-sector partners, as appropriate, should launch and conduct the lunar surface power demonstration project.

(c) By the late-2020s, establish the technical foundations and capabilities — including through identification and resolution of the key technical challenges — that will enable NTP options to meet future DoD and NASA mission needs.

(i) Conduct requirements assessment. DoD and NASA, in cooperation with DOE, and with other agencies and private-sector partners, as appropriate, should assess the ability of NTP capabilities to enable and advance existing and potential future DoD and NASA mission requirements.

(ii) Conduct technology assessment. DoD and NASA, in cooperation with DOE, and with other agencies and private-sector partners, as appropriate, should evaluate technology options and associated key technical challenges for an NTP system, including reactor designs, power conversion, and thermal management. DoD and NASA should work with their partners to evaluate and use opportunities for commonality with other SNPP needs, terrestrial power needs, and reactor demonstration projects planned by agencies and the private sector.

(iii) Technology development. DoD, in coordination with DOE and other agencies, and with private-sector partners, as appropriate, should develop reactor and propulsion system technologies that will resolve the key technical challenges in areas such as reactor design and production, propulsion system and spacecraft design, and SNPP system integration.

(d) By 2030, develop advanced RPS capabilities that provide higher fuel efficiency, higher specific energy, and longer operational lifetime than existing RPS capabilities, thus enabling survivable surface elements to support robotic and human exploration of the Moon and Mars and extending robotic exploration of the solar system.

(i) Maintain RPS capability. Mission sponsoring agencies should assess their needs for radioisotope heat source material to meet emerging mission requirements, and should work with DOE to jointly identify the means to produce or acquire the necessary material on a timeline that meets mission requirements.

(ii) Engage the private sector. NASA, in coordination with DOE and DOC, should conduct an assessment of opportunities for engaging the private sector to meet RPS needs in an effective manner consistent with the guiding principles established in this memorandum.

(iii) Conduct technology and requirements assessment. NASA, in coordination with DOE and DoD, and with other agencies and private-sector partners, as appropriate, should assess requirements for next-generation RPS systems and evaluate technology options for meeting those requirements.

(iv) System development. DOE, in coordination with NASA and DoD, and with other agencies and private-sector partners, as appropriate, should develop one or more next-generation RPS system or systems to meet the goals of higher fuel efficiency, higher specific energy, and longer operational lifetime for the required range of power.

Sec. 6. Implementation. The Vice President, through the National Space Council, shall coordinate implementation of this memorandum.

Sec. 7. General Provisions. (a) Nothing in this memorandum shall be construed to impair or otherwise affect:

(i) the authority granted by law to an executive department or agency, or the head thereof; or

(ii) the functions of the Director of the Office of Management and Budget relating to budgetary, administrative, or legislative proposals.

(b) This memorandum shall be implemented consistent with applicable law and subject to the availability of appropriations.

(c) This memorandum is not intended to, and does not, create any right or benefit, substantive or procedural, enforceable at law or in equity by any party against the United States, its departments, agencies, or entities, its officers, employees, or agents, or any other person.

(d) The Secretary of Energy is authorized and directed to publish this memorandum in the Federal Register.

DONALD J. TRUMP,
The White House

Appendix 2

Executive Summary of the National Academies Report on Space Nuclear Propulsion for Human Mars Exploration (2021)

In 2020, the National Academies of Sciences, Engineering, and Medicine convened the ad hoc Space Nuclear Propulsion Technologies Committee to identify primary technical and programmatic challenges, merits, and risks for maturing space nuclear propulsion technologies of interest to a future human Mars exploration mission. Through interactions with experts from across the space propulsion community, the committee assessed the present state of the art, a potential development path, and key risks for (1) a nuclear thermal propulsion (NTP) system designed to produce a specific impulse of at least 900 s and (2) a nuclear electric propulsion (NEP) system with at least one megawatt of electric (MWe) power and a mass-to-power ratio that is substantially lower than the current state of the art. As requested by NASA, each system was assessed with regard to its ability to support a particular baseline mission—an opposition-class human exploration mission to Mars with a 2039 launch date.

For both NEP and NTP systems, efforts to mature the requisite technology and mitigate key technical risks were integrated into a top-level development and demonstration roadmap. Infusion of technology results, expertise, and synergy with other government programs and missions was also examined. In the near term, NASA and the Department of Energy (DOE), with inputs from other key stakeholders, including commercial industry and academia, should conduct a comprehensive assessment of the relative merits and challenges of highly enriched uranium (HEU) and high-assay, low-enriched uranium (HALEU) fuels for NTP and NEP systems as applied to the baseline mission.

For NEP systems, the fundamental challenge is to scale up the operating power of each NEP subsystem and to develop an integrated NEP system suitable for the baseline mission. This requires, for example, scaling power and thermal management systems to power levels orders of magnitude higher than have been achieved to date. While no integrated system testing has ever been performed on MWe-class NEP systems, operational reliability over a period of years is required for the baseline mission. Lastly, application of a complex set of NEP subsystems to the baseline mission requires parallel development of a compatible large-scale chemical propulsion system to provide the primary thrust when departing Earth orbit and when entering and departing Mars orbit. As a result of low and intermittent investment over the past several decades, it is unclear if even an aggressive program would be able to develop an NEP system capable of executing the baseline mission in 2039.

NTP development faces four major challenges that an aggressive program could overcome to achieve the baseline mission in 2039. The fundamental challenge is to develop an NTP system that can heat its propellant to approximately 2700 K at the reactor exit for the duration of each burn. The other three challenges are the long term storage of liquid hydrogen in space with minimal loss, the lack of adequate ground-based test facilities, and the need to rapidly bring an NTP system to full operating temperature (preferably in 1 min or less). Although the United States has conducted ground-based testing of NTP technologies, those tests took place nearly 50 years ago, and did not fully address flight system requirements: recapturing the ability to conduct necessary ground testing will be costly and time-consuming. Furthermore, no in-space NTP system has ever been operated.

Despite recent work in fuel development, this area remains a challenge, particularly for NTP systems. A comprehensive assessment of HALEU versus HEU for NTP and NEP systems that evaluates a full set of critical parameters as applied to the baseline Mars mission has not been performed. Similarly, a recent apples-to-apples trade study comparing NEP and NTP systems for crewed missions to Mars, in general, or the baseline mission in particular does not exist. The committee recommends that NASA and DOE, with inputs from other key stakeholders, including commercial industry and academia, conduct a comprehensive and expeditious assessment of the relative merits and challenges of HEU and HALEU fuels for NTP and NEP systems as applied to the baseline mission.

The committee recommends that the development of operational NTP and NEP systems include extensive investments in modeling and simulation. Ground and flight qualification testing will also be required. For NTP systems, ground testing should include integrated system tests at full scale and thrust. For NEP systems, ground testing should include modular subsystem tests at full scale and power. Given the need to send multiple cargo missions to Mars prior to the flight of the first crewed mission, the committee also recommends that NASA use these cargo missions as a means of flight qualification of the space nuclear propulsion system that will be incorporated into the first crewed mission.

NEP and NTP systems show great potential to facilitate the human exploration of Mars. Using either system to execute the baseline mission by 2039, however, will require an aggressive research and development program. Such a program would need to begin with NASA making a significant set of architecture and investment decisions in the coming year. In particular, NASA should develop consistent figures of merit and technical expertise to allow for an objective comparison of the ability of NEP and NTP systems to meet requirements for a 2039 launch of the baseline mission.

The committee recommends that the development of operational NTP and NEP systems include extensive investments in modeling and simulation. Ground and flight qualification testing will also be required. For NTP systems, ground testing should include integrated system tests at full scale and thrust. For NEP systems, ground testing should include modular subsystem tests at full scale and power. Given the need to send multiple cargo missions to Mars prior to the flight of the first crewed mission, the committee also recommends that NASA use these cargo missions as a means of flight qualification of the space nuclear propulsion system that will be incorporated into the first crewed mission.

NEP and NTP systems show great potential to facilitate the human exploration of Mars. Using either system to execute the baseline mission by 2039, however, will require an aggressive research and development program.

Such a program would need to begin with NASA making a significant set of architecture and investment decisions in the coming year. In particular, NASA should develop consistent figures of merit and technical expertise to allow for an objective comparison of the ability of NEP and NTP systems to meet requirements for a 2039 launch of the baseline mission.

SUMMARY

NASA is presently considering multiple forms of propulsion, including NTP and NEP, in its mission architecture analyses. Opposition-class missions, while reducing crew duration on Mars and total mission time, markedly increase mission ΔV requirements. This mission class introduces a higher sensitivity in propulsion system requirements from one launch opportunity to another, which could be achieved by either an NTP or NEP system.

Successful development of an NTP or NEP/chemical system at relevant scale and performance would allow NASA to develop a robust architecture with flexibility across multiple mission opportunities.

This report provides a technology assessment of the NTP and NEP development challenges that must be overcome to execute the baseline Mars mission. It is not intended to provide—nor did the committee's statement of task allow—a comprehensive assessment of all aspects or trade studies associated with how a human Mars exploration mission should be organized, funded, or executed.

Appendix 3

Team Biographies

Counting the scientists, engineers, management and support personnel working on the projects described in this book, there must be over a thousand people involved counting NASA, DARPA, DOD's National Laboratories and all the supporting contractors. This appendix could not possibly include all of their biographies but included are those that are often quoted, seen in videos or are in key positions. My apologies to those not included but know that your contributions are duly noted. You should take great satisfaction in having supported projects that may lead to a new way to travel in deep space. The following people are listed in alphabetical order within each category.

A3.1 NASA Headquarters

Fig. A3.1 Dr. Jacob Bleacher
Photo courtesy of NASA.

Dr. Jacob Bleacher is the Chief Exploration Scientist for NASA's Human Exploration and Operations Mission Directorate (HEOMD) at NASA Headquarters. He is the science advocate for NASA technology and architecture development that is intended to enable human exploration of the Moon and Mars. He also serves as a primary contact with NASA's Science Mission Directorate (SMD) and the science community external to NASA.

Dr. Bleacher earned a B.S. in Geosciences from Franklin and Marshall College and Ph.D. in Geological Sciences from the Arizona State University. During his Ph.D. research he worked on the European Space Agency's Mars Express Mission by conducting geologic mapping of the initial images acquired of the large Tharsis province volcanoes by the Mission's High Resolution Stereo Camera (HRSC). Dr. Bleacher joined NASA's Goddard Space Flight Center (GSFC) as a NASA Postdoctoral Program Fellow after which he was hired by NASA as a research scientist.

Bleacher's research focuses on understanding the volcanic history of the Earth, Moon, and Mars by remote sensing mapping and field work. Upon joining the NASA workforce he began supporting the Constellation Program Office to conduct studies examining potential landing sites and developing science traverse plans to help define requirements for hardware on the lunar surface. Dr. Bleacher served as a science team member and test subject as a crew member for NASA's Desert Research and Technology Studies (Desert RATS), which focused on field tests of operations associated with prototype human rovers for the Moon. He has served as the lead for the Goddard Instrument Field team, in which he organized GSFC's planetary science field research deployments as well as GSFC's Lead Exploration Scientist, for which he worked to create cross-Center awareness of ongoing human exploration projects.

Bleacher is a co-author on a number of peer reviewed science publications including a cover article for the journal Nature focused on explosive volcanism on Mars as well as the chapter on Mars Volcanism in the Encyclopedia of Volcanoes, 2nd edition. He has served on a series of Special Action Teams (SATs) for the Lunar Exploration Analysis Group (LEAG) and Mars Exploration Program Analysis Group (MEPAG) including the Lunar Sample Acquisition and Curation Review, Candidate Scientific Objectives for the

Human Exploration of Mars, and Geological Astronaut Training and has served on NASA's In Situ Resource Utilization (ISRU) Systems Capability Leadership Team.

Fig. A3.2 Dr. Anthony Calomino
Photo courtesy of CU Boulder.

Dr. Anthony M. Calomino manages the Space Nuclear Power and Propulsion technologies under the NASA Space Technology Mission Directorate (STMD). His focus is the advancement of space nuclear fission power and propulsion capabilities to meet near-term human exploration missions to the Moo n and Mars, as well as for future robotic missions. This involves durability analysis and damage modeling of high-temperature materials and composites extending from metallic super alloys, to ceramic matrix composites, ablators, and refractory soft goods. He has worked in nuclear fission technology development since 2016.

Previously, Dr. Calomino was the NASA Materials technical lead for Entry Systems Modeling project and the Deputy Principal Investigator for Flexible Systems Development under NASA's Hypersonic Inflatable Aerodynamic Decelerator Programs.

He has a B.S. and M.S in Structural and Engineering Mechanics, and a Ph.D. in Materials Science from Northwestern University.

Fig. A3.3 Pam Meloy
Photo courtesy of NASA.

Col. (USAF, ret) Pam Melroy was sworn in as the NASA Deputy Administrator in 2021. Melroy performs the duties and exercises the powers delegated by the Administrator, assists the Administrator in making final agency decisions, and acts for the Administrator in his absence by performing all necessary functions to govern NASA operations. Melroy is also responsible for laying the agency's vision and representing NASA to the Executive Office of the President, Congress, heads of federal and other appropriate government agencies, international organizations, and external organizations and communities.

Melroy was commissioned through the Air Force Reserve Officers' Training Corps (ROTC) program in 1983. As a co-pilot, aircraft commander, instructor pilot, and test pilot, Melroy logged more than 6,000 flight hours in more than 50 different aircraft before retiring from the Air Force in 2007. She is a veteran of Operation Desert Shield/Desert Storm and Operation Just Cause, with more than 200 combat and combat support hours.

Melroy was selected as an astronaut candidate by NASA in December 1994. Initially assigned to astronaut support duties for launch and landing, she also worked advanced projects for the Astronaut Office. She also performed Capsule Communicator (CAPCOM) duties in Mission Control. In addition, she served on the Columbia Reconstruction Team as the lead for the crew module and served as Deputy Project Manager for the Columbia Crew Survival

Investigation Team. In her final position, she served as Branch Chief for the Orion branch of the Astronaut Office.

One of only two women to command a Space Shuttle, Melroy logged more than 38 days (924 hours) in space. She served as pilot on two flights, STS-92 in 2000 and STS-112 in 2002, and was the mission commander on STS-120 in 2007. All three of her missions were assembly missions to build the International Space Station.

After serving more than two decades in the Air Force and as a NASA astronaut, Melroy took on a number of leadership roles, including at Lockheed Martin, the Federal Aviation Administration, the Defense Advanced Research Projects Agency, Nova Systems Pty, Australia, and as an advisor to the Australian Space Agency. She also served as an independent consultant and a member of the National Space Council's Users Advisory Group.

Melroy holds a B.S. Physics and Astronomy from Wellesley College and a M.S. degree in Earth and Planetary Sciences from the Massachusetts Institute of Technology.

Fig. A3.4 James. L. Reuter.
Photo courtesy of NASA.

James L. Reuter was named NASA's Associate Administrator for the Space Technology Mission Directorate (STMD) at NASA Headquarters in June

2019, a position in which he served in an acting capacity since February 2018. In this role, he provides executive leadership and management of the technology programs within STMD, with an annual investment value of more than $1 billion.

Reuter was the Deputy Associate Administrator of STMD from February 2017-February 2018. Prior to this role, Reuter served as the senior executive for technical integration in the Center Director's Office at NASA's Marshall Space Flight Center in Huntsville, Alabama, from 2009-2015, providing strategic leadership on critical technology and integration activities. Additionally, Reuter served as the Exploration Systems Division (ESD) standing review board chair, responsible for overseeing development activities of the Space Launch System, Orion Multi-Purpose Crew Vehicle, Ground Systems Development and Operations Programs, and the ESD integration activities.

Previously, Reuter served in many managerial roles at Marshall including Ares vehicle integration manager in the Constellation program, the Deputy Manager of Space Shuttle Propulsion Office, and the Deputy Manager of Space Shuttle External Tank Project Office during the shuttle return-to-flight activities. In 2002, he was assigned to a detail at NASA Headquarters as the Deputy Associate Director in the Space Transportation Technology Division in the Office of Aerospace Technology. From 1994 to 2001, he was the Environmental Control and Life Support System manager for the International Space Station at NASA's Johnson Space Center. Reuter began his NASA career in 1983 as an aerospace engineer in the Structures and Propulsion Laboratory in Marshall's Science and Engineering Directorate.

Reuter has a B.S degree in Mechanical Engineering from the University of Minnesota in Minneapolis. He has received numerous NASA awards and honors, including a 2019 Distinguished Service Medal, 2016 Outstanding Leadership Medal, 2013 NASA Exceptional Achievement Medal, a 2008 NASA Outstanding Leadership Medal, and a 2002 NASA Exceptional Service

A3.2 NASA CENTERS

Fig. A3.5 Dr. Stanley K. Borowski
Photo courtesy of AIAA.

Dr. Stan Borowski is a senior aerospace / nuclear engineer and branch chief of the Propulsion & Controls Systems Analysis group at NASA's Glenn Research Center. During his past 21 years, he has been GRC's technical lead for all human and robotic space transfer vehicle design and analysis activities involving the use of Nuclear Thermal Rocket (NTR) propulsion for exploration missions to the Moon, Mars, near Earth asteroids (NEAs) and the outer planets. His latest responsibilities in this area included leading the NTR space transportation analysis team during NASA's recent Mars Design Reference Architecture (5.0) study.

He received his B.S. and M.S. degrees in Nuclear Engineering from the Pennsylvania State University, and his Ph.D. in Nuclear Engineering from the University of Michigan.

Fig. A3.6 Dr. William J. Emrich

Dr. Emrich is the NASA Marshall Space Flight Center Project Manager for the megawatt-class Nuclear Thermal Rocket Element Environment Simulator, or NTREES. He conceived, designed and brought the simulator to operational status. For that effort, the American Institute of Aeronautics and Astronautics honored Bill Emrich with its prestigious 2015 Engineer of the Year award.

NTREES allows engineers and researchers to perform realistic, non-nuclear testing of prototypical nuclear rocket fuel elements by creating an environment that simultaneously reproduces the power, flow and temperature conditions that the fuel element would be expected to encounter during actual nuclear engine operation.

Now retired from NASA Marshal after 35 years, he is currently an adjunct professor at the University of Alabama in Huntsville where he teaches a course in nuclear rocket propulsion and mentors young engineers seeking to pursue a career in that field. William Emrich is a registered Professional Engineer in the state of California and is a fellow in the American Society Mechanical Engineers and an associate fellow in the American Institute of Aeronautics and Astronautics.

Dr. Emrich earned a B.S. in Mechanical Engineering from the Georgia Institute of Technology; a M.S. degree in Nuclear Engineering from the Massachusetts Institute of Technology; and a Ph.D. in Mechanical and Aerospace Engineering from the University of Alabama.

Fig.A3.7 Marc Gibson
Photo courtesy of NASA.

Marc Gibson is the Glenn Research Center's lead engineer for NASA's nuclear systems Kilopower project tasked with advancing the technology readiness of fission power systems for space.

Marc started his career at NASA in 2007 after working in the private sector for ten years as chief engineer for numerous commercial and government research projects. Since being at NASA, Marc has been responsible for the engineering and development of nuclear systems for in-space and planetary surface power in support of the Space Technology Mission Directorate.

Marc received a B.S.in Mechanical Engineering from the University of Akron, Ohio and a M.S. in Aerospace Engineering from the Case Western Reserve University.

Fig. A3.8 Michael G. Houts
Photo courtesy of NASA.

Dr. Houts is the Nuclear Research Manager at the NASA Marshall Space
Flight Center. He has over 20 years of experience in the field of nuclear
engineering, specializing in space nuclear power and propulsion. Dr. Houts
holds B.S. degrees in Mechanical Engineering and Nuclear Engineering from
the University of Florida, as well as a Ph.D. in Nuclear Engineering from the
Massachusetts Institute of Technology.

Dr. Houts worked at Los Alamos National Laboratory for 11 years, where he
held a variety of positions including Team Leader for Criticality, Reactor, and
Radiation Physics and Deputy Group Leader of the 70 person Nuclear Design
and Risk Analysis group. Dr. Houts has been employed by NASA for 9 years.

He is a member of the American Nuclear Society and the American Institute
of Aeronautics and Astronautics. Dr. Houts has authored papers on a variety of
subjects, including space nuclear power and propulsion, actinide transmutation,
tritium production, radiation shielding, and others. Dr. Houts currently serves
on NASA's Nuclear Systems Working Group and the Radioisotope Power
Systems Standing Review Board.

He has been the lead author on more than two dozen technical papers related
to space nuclear power and propulsion, as well as additional papers related to
Accelerator Transmutation of Waste, Plutonium Disposition, and other topics.
He has co-authored over 100 technical papers and has received numerous
distinguished performance awards, including a NASA Exceptional Service

Medal, a NASA Exceptional Engineering Achievement Medal, and was selected as an Associate Fellow of the American Institute of Aeronautics and Astronautics. In 2020, he received NASA's most prestigious and distinguished honor, the Distinguished Service Medal. In 2020, Dr. Houts received NASA's most prestigious and distinguished honor, the "Distinguished Service Medal." His work involving the use of nuclear reactors in space, moving America towards visiting Mars, led to his being quoted by Scientific American in 2022. For a 1:00 minute video of Dr. Houts, go to: https://youtu.be/COrA5VjeviU

Fig. A3.9 Sonny Mitchell
Photo courtesy of NASA.

Prior to becoming the Program Element Manager of the Game Changing Development Program (GCD), Sonny Mitchell served as Project Manager for the Nuclear Thermal Propulsion project at NASA's Marshall Space Flight Center. In this role, he led a team in the development of a low-enriched uranium nuclear rocket engine for human Mars missions. During his career, Sonny also served as Chief for the Systems Engineering and Integration Division and as Chief Engineer for several flight projects.

Sonny holds a B.S. degree in Mechanical Engineering and has more than 32 years of experience developing and flying space flight systems at NASA. His areas of expertise include project management, systems engineering, and mission operations.

Fig. A3.10 Michelle Rucker
Photo courtesy of AIAA

Ms. Rucker currently leads the Mars Architecture Team for the Human Exploration and Operations Mission Directorate's System Engineering and Integration (SE&I) Office at Kennedy Space Center. A native of Anchorage, Alaska and a 35-year veteran of NASA, she began her engineering career in the Houston oil industry, designing down-hole sensors while pursuing undergraduate and graduate degrees in Mechanical Engineering from Rice University. She joined NASA in the aftermath of the Space Shuttle Challenger accident, supporting the investigation team by conducting booster material test and analysis at NASA's White Sands Test Facility. She has worked on a wide array of projects, ranging from hypervelocity impact research to spacesuit and Extravehicular Activity (EVA) tools development, microgravity exercise equipment system engineering, and lunar lander test and verification.

At White Sands, she also managed the two-stage light gas gun hypervelocity impact research laboratory and developed environmental control and life support systems for the International Space Station (ISS) verification.

Fig. A3.11 Dr. George Williams
Photo courtesy of NASA/GRC

Dr. Williams is a Research Aerospace Scientist and Technologist currently involved in the development of plasma and chemical propulsion systems at NASA GRC where he has been since 2017. He is a part time lecturer at Cleveland State University. I have enjoyed mentoring both graduate and undergraduate students, and teach courses in rocket propulsion and applied heat transfer.

His work includes:

- Mission support studies for Ion Propulsion systems including Mars sample return mission, outer planet exploration, and near-Earth commercial applications
- Design, service-life modeling, and performance and wear testing of ion and Hall-effect thrusters.
- Test coordination, testing and performance modeling of LOX-Methane rockets
- Development and implementation of various lasers and other optical diagnostics including LIF, interferometry, OES, and surface imaging.

From September, 2000 to January, 2017, Dr. Williams was a Senior Scientist At the Ohio Aerospace Institute/ NASA GRC. He has a Ph.D. in Aerospace Engineering from the University of Michigan, an M.S. in Engineering Applied

Physics, Mechanical and Aerospace Engineering from Princeton University. His BAE, M.S. in Aerospace Engineering is from Auburn University.

A3.3 DARPA

Fig. A3.12 Dr. Tabitha Dodson
Photo courtesy of DARPA

Dr. Tabitha Dodson joined the DARPA Tactical Technology Office as a program manager in August 2021. Her interests are in: advanced space payloads, electric propulsion, astrodynamics, nuclear thermal and nuclear electric propulsion, overall rocket propulsion, advanced nuclear reactors, plasma physics and plasma engineering, nuclear/quantum/particle physics, and hypersonics.

Dr. Dodson is the Program Manager for the Demonstration Rocket for Agile Cislunar Operations (DRACO) program. Prior to becoming a program manager at DARPA, Dodson was a Systems Engineering and Technical Assistance (SETA) contractor with Gryphon-Schafer Government Services, LLC, also within DARPA, where she served beginning in 2018 as the chief engineer of the DRACO program, which endeavors to build and test a nuclear thermal rocket.

Upon conversion to being a government employee in 2021, she continued to serve as DRACO's chief engineer, concurrently as its deputy program manager. Dodson has worked as adjunct professor of aerospace engineering at the United States Naval Academy and also as an adjunct professor in the Aeronautics and Astronautics Department of the Air Force Institute of Technology (AFIT). Dodson has worked in various other positions within and for the U.S. Air Force, including as an aerospace engineer and senior scientist in the fields of spacecraft engineering, space missions and operations, space power, and space propulsion. She was also a Graduate Student Researchers Program fellow with NASA conducting research in materials for nuclear thermal propulsion.

Dodson holds a doctorate in applied physics from the Air Force Institute of Technology, as well as a Ph.D. in mechanical and aerospace engineering from the George Washington University (GWU), where she also earned a master's in space policy, in addition to bachelor's degrees in physics, sociology, and anthropology.

Fig. A3.13 Dr. Stefanie Tompkins
Photo courtesy of NASA.

Dr. Stefanie Tompkins is the director of the Defense Advanced Research Projects Agency (DARPA). Prior to this assignment, she was the vice president for research and technology transfer at Colorado School of Mines.

Tompkins has spent much of her professional life leading scientists and engineers in developing new technology capabilities. She began her industry career as a senior scientist and later assistant vice-president and line manager at Science Applications International Corporation, where she spent 10 years conducting and managing research projects in planetary mapping, geology, and imaging spectroscopy. As a program manager in DARPA's Strategic Technology Office, she created and managed programs in ubiquitous GPS-free navigation as well as in optical component manufacturing. Tompkins has also served as the deputy director of DARPA's Strategic Technology Office, director of DARPA's Defense Sciences Office; the agency's most exploratory office in identifying and accelerating breakthrough technologies for national security as well as the Acting DARPA Deputy Director.

Tompkins received a Bachelor of Arts degree in geology and geophysics from Princeton University and Master of Science and Doctor of Philosophy degrees in geology from Brown University. She has also served as a military intelligence officer in the U.S. Army.

A3.4 DOE NATIONAL LABORATORIES

Fig. A3.14 Dr. Stephen Johnson
Photo courtesy of DOD/INL

Dr. Stephen G. Johnson is the Director of the Space Nuclear Power and Isotope Technologies Division at the Idaho National Laboratory's Materials and Fuels Complex.

NASA is working with the Idaho National Laboratory (INL) on providing enough power to establish an outpost on the Moon or Mars. The ability to produce large amounts of electrical power on planetary surfaces using a fission surface power system that would enable large-scale exploration, establishment of human outposts, and utilization of in situ resources, while allowing for the possibility of commercialization. Dr. Johnson believes that the on-going efforts will be able to deliver a power system for the Artemis Program.

During Johnson's tenure, the laboratory has successfully pursued involvement in the Radioisotope Power Systems program. Following that involvement, the fueling and testing of space and terrestrial power systems operations. He initiated the Technical Integration Office for the DOE's national Space and Defense Power Systems (SPDS) program. This office focuses on cross-cutting initiatives, mission planning scenarios for nuclear power systems for DOE customers such as NASA, and provides resources for various other needs for DOE's SDPS department.

Johnson holds a B.S. degree with a double major in Mathematics and Chemistry from Lake Superior State University of Michigan and a Ph. D. in Physical Chemistry from Iowa State University.

Fig. A3.15 Michael B. Smith
Photo courtesy of ORNL

Mr. Smith joined ORNL in 2017 in a post-Masters appointment where he worked in the Advanced Reactor Engineering Group supporting analysis for NASA's Radioisotope Power System Program and Nuclear Thermal Propulsion initiatives. Michael transitioned to the staff at the ORNL in 2018 where he continued to support analysis and research for the space program.

Michael currently works in the Systems Analysis and Optimization Group where he leads and supports multiple research and analysis activities with a focus on radiation modeling of spaceflight scenarios. These efforts primarily support NASA, Department of Defense, and industry activities to estimate and predict deleterious radiation environments.

Michael B. R. Smith received his B.S. and M.S. degrees in Nuclear Engineering from the University of Tennessee. His research focused on galactic cosmic ray effects on the Martian surface and radioisotope power system (RPS) radiation environments. During his studies, Michael worked as an intern at the Oak Ridge National Laboratory (ORNL) at the Spallation

Neutron Source (SNS) on the Personal Protection Systems team developing, authoring, and testing safety procedures for neutron beamlines before transitioning to the Neutronics team where he supported beamline diagnostics and facility-wide radiation background analysis.

In his previous life, Mr. Smith was a diver and has a USCG Manned Submersible endorsement. He has logged over 1,000 subsurface hours as master (pilot), and 3,000 subsurface hours as crew (co-pilot) in the Atlantis IX and IV vessels off the coast of Maui, HI

Fig. A3.16 John C. Wagner
Photo courtesy of INL.

Dr. John C. Wagner is the laboratory director for Idaho National Laboratory. His previous roles included Associate Laboratory Director of INL's Nuclear Science & Technology (NS&T) Directorate, director of Domestic Programs in NS&T as well as director of the Technical Integration Office for the DOE-NE Light Water Reactor Sustainability Program at INL. Wagner initially joined INL as the chief scientist at the Materials and Fuels Complex in 2016. He has more than 20 years of experience performing research, and managing and leading research and development projects, programs and organizations.

Wagner received a B.S. in nuclear engineering from the Missouri University of Science and Technology in 1992, and M.S. and Ph.D. degrees from the

Pennsylvania State University in 1994 and 1997, respectively. Following graduate school, Wagner joined Holtec International as a principal engineer, performing criticality safety analyses and licensing activities for spent fuel storage pools and storage and transportation casks. Wagner joined the Oak Ridge National Laboratory (ORNL) as an R&D staff member in 1999, performing research in the areas of hybrid (Monte Carlo/deterministic) radiation transport methods, burnup credit criticality safety, and spent nuclear fuel characterization and safety.

While at ORNL, Wagner held various technical leadership positions, including technical lead for post closure criticality in support of DOE OCRWM's Lead Laboratory for Repository Systems, Radiation Transport Methods Deputy Focus Area lead for the Consortium for Advanced Simulation of Light Water Reactors (CASL), and national technical director of the DOE Office of Nuclear Energy's Nuclear Fuels Storage and Transportation Planning Project. Wagner also held various management positions, including group leader for the Criticality and Shielding Methods and Applications, Radiation Transport, and Used Fuel Systems groups.

In 2014, Wagner became director of the Reactor and Nuclear Systems Division (RNSD), with responsibility for management direction and leadership to focus and integrate the seven RNSD R&D groups and the Radiation Safety Information Computational Center.

Wagner is a Fellow of the American Nuclear Society and recipient of the 2013 E.O. Lawrence Award. He has authored or co-authored more than 170 refereed journal and conference articles, technical reports, and conference summaries.

A3.5 AEROSPACE INDUSTRY

Fig. A3.17 Dr. Michael Eades
Photo courtesy of USNC.

Dr. Eades is the Chief Engineer at of the Tech Division at USNC and provides technical leadership on key projects. For more than a decade, he has been integral to the advancement of modern space nuclear systems and has authored more than 30 combined conference papers and journal articles on space nuclear topics. He has been involved in almost every NASA design reactor development effort since 2014 and was instrumental in establishing HALEU fueled space nuclear technology as a field of development.

Before USNC, he was previously a visiting technologist at NASA Marshal Space Flight Center (MSFC) and Center for Space Nuclear Research (CSNR) at the Idaho National Laboratory (INL). While at USNC-Tech, Dr. Eades has lead projects from external customers such as Blue Origin, BWXT, and Acrojet-Rocketdyne.

He started at USNC in 2016 and was the second engineer on staff. He guided the rapid growth of USNC-Tech in 2020-2021 from 7 people to 41 people as Director of Engineering. At USNC, he initiated and set technical direction for USNC's HALEU NTP systems, EmberCore, and Pylon reactor products. He has represented USNC to media outlets including CNN, IEEE Spectrum, and The Economist.

Fig. A3.18 David Durham
Photo courtesy of Westinghouse

David Durham is President, Energy Systems at Westinghouse Electric Company. In this role, he leads the global business for Westinghouse AP1000® reactors, small modular reactors and micro reactors, as well as advanced reactor development and other non-nuclear energy systems.

Mr. Durham joined Westinghouse in 2015, as Senior Vice President, New Projects Business, and President of WECTEC LLC, a wholly-owned Westinghouse subsidiary that provides engineering and construction support services to AP1000 plant projects in China and the U.S., as well as government services and contingency staffing services. He assumed responsibility for Westinghouse's AP1000 plant projects and Decommissioning business lines early in 2017, serving as President, Plant Solutions Business.

Prior to Westinghouse, Mr. Durham was Senior Vice President and Chief Commercial Officer at GE Hitachi Nuclear Energy. Previously, he was Vice President of Fluor Corporation, leading commercial activities for the company's global nuclear power and government businesses. Earlier in his career, Mr. Durham served in the administration of U.S. President George H.W. Bush, overseeing the creation and operations of the defense nuclear facilities decommissioning and cleanup program for Energy Secretary James Watkins, as well as federal facilities enforcement for U.S. Environmental Protection Agency Administrator Bill Reilly. Mr. Durham holds a B.S. degree in foreign affairs and history from Assumption College and a Juris Doctor degree from the George Washington University National Law Center.

Fig. A3.19 Dr. Pete Pappano
Photo courtesy of X-energy

As President of TRISO-X, Dr. Pete Pappano leads X-energy's fuel production strategy. Since joining X-energy in 2015, he has managed the company's low enriched uranium supply chain and the development of X-energy's fuel fabrication capabilities. Dr. Pappano leads X-energy's DOE ARC15 project, which enables X-energy to further its reactor design codes and methods, research/develop a TRISO-based fuel production capability, and initiate interactions with the Nuclear Regulatory Commission (NRC). As Principal Investigator for the DOE iFOA project, he manages X-energy's team that will complete the design of a CAT II facility and develop an NRC fuel fabrication license application.

Dr. Pappano has 20 years of experience in the public and private sector, including the U.S. Department of Energy, Oak Ridge National Laboratory and

SGL Carbon, where he has managed nuclear energy projects and multiple cooperative research and development agreements with DOE offices. Pete was awarded undergraduate and master's degrees from Penn State University and went on to earn a PhD in Materials Science from the University.

Fig. A3.20 Dr. Brad Rearden
Photo courtesy of X-energy.

Dr. Brad Rearden is the Director of the Government R&D Division at X-energy, providing innovative nuclear energy solutions, especially through the development of transportable high-temperature micro reactors, with a focus on supporting the missions of government agencies including DOD, DOE, NASA, and others. Dr. Rearden brings more than 23 years of experience as a values-driven, hands-on leader providing innovative solutions and analysis approaches that have enabled advances in the design, deployment, and regulation of nuclear technologies, implementing successful outcome-oriented practices focused on quality assurance and team building in the realization of world-class capabilities.

He served as Deputy Program Manager and Director of Engineering in X-energy's development of a megawatt class, containerized, transportable nuclear power system for DOD's Project Pele and as Solution Architect for X-energy's space nuclear programs including Lunar Fission Surface Power systems and space nuclear propulsion concepts.

During his 20-year tenure at ORNL, he led Modeling and Simulation Integration in the Reactor and Nuclear Systems Division, where he inspired innovation and efficiency through ORNL's Nuclear Resources Analysis and Modeling Portfolio (ONRAMP) programs. Brad has served as technical or programmatic lead for $100M of applications-driven research, development, deployment, and training projects, for the U.S. Nuclear Regulatory Commission (NRC), the U.S. Department of Energy (DOE), and other domestic and international industrial partners.

As the Director of the SCALE Code System from 2009-2018, he developed and fulfilled a progressive multi-year application-driven strategy to establish one of the world's premiere toolkits for nuclear design, analysis and regulatory review, licensed to 10,000 users in over 60 nations. Brad also served as the leader of the Integration Product Line for the Nuclear Energy Advanced Modeling and Simulation program from 2015-2018 and as the National Technical Director of the DOE Office of Nuclear Energy's Nuclear Data and Benchmarking Program in 2018.

Dr. Rearden is the author or co-author of more than 200 publications including peer reviewed journals, technical reports, and conference papers. He has held leadership roles in professional societies and international working groups and is the recipient of numerous awards for technical excellence, technology transfer, and mentoring. He holds B.S., M.S., and Ph.D. degrees in Nuclear Engineering from Texas A&M University.

Fig. A3.21 Dr. Paolo Venneri

Dr. Paolo Venneri is the CEO and founder of USNC-Tech and Executive Vice President, Advanced Technologies Division. He is responsible for setting company direction, managing multiple projects, and building new business and growth opportunities. He has overseen multiple successful Phase I and Phase II NASA SBIRs and leads projects related to developing fuel and reactor designs for nuclear thermal propulsion, nuclear electric propulsion, and surface fission power.

His technical background includes researching the development of thermal spectrum Low-Enriched Uranium Nuclear Thermal Propulsion (LEU-NTP) systems for the past five years. He was the first to publish on LEU-NTP systems' neutronic design and is involved in the development of the LEU graphite composite, tungsten cermet, and various advanced NTP systems.

Paolo Venneri earned a Ph.D. from the Korea Advanced Institute of Science and Technology in Nuclear and Quantum Engineering with the thesis of determining the feasibility of low enriched uranium for nuclear thermal propulsion. At USNC, he founded and led the Advanced Projects Division in design efforts to support the NASA LEU-NTP program as well as being a major contributor to USNC's innovative nuclear power designs developed for remote community and mining sites. He was the first to propose and computationally demonstrate the feasibility of HALEU fuel in nuclear thermal propulsion and has been a driver for the continued technical development of commercially viable space nuclear systems. In January 2019, he led the splitting of USNC-Tech as a separate and independently managed subsidiary of USNC to focus on development of advanced nuclear technology and space nuclear systems.

Fig. A3.22 Dr. Jonathan Witter
Photo courtesy of BWXT

Jonathan Witter is the Chief Engineer for BWXT's Advance Technology Programs. In this role he currently serves as the technical engineering lead for the NASA GCD Nuclear Thermal Propulsion (NTP) project with a focus on the reactor core design and analysis and the fuel mechanical development and testing.

Dr. Witter has past experience with space nuclear programs where he served as a reactor physics design lead for the Project Prometheus/Jupiter Icy Moon Orbiter fission power system in the early 2000's while working at the Knolls Atomic Power Laboratory and did his Ph.D. work at Massachusetts Institute of Technology under a NASA Space Grant working on nuclear thermal propulsion under the Space Exploration Initiative in the early 1990's.

He received his B.S. (1982) and Masters (1983) in Nuclear Engineering from Rensselaer Polytechnic Institute. After that he spent 6 years in nuclear industry learning the operations of naval propulsion and commercial nuclear power plants. Wanting to realize the goal of getting his Ph.D. after physical operational experience, Jonathan went to MIT and dove in head-first into the unique propulsion system of a nuclear rocket engine and graduated from the Nuclear Engineering Department in 1993. After MIT, Dr. Witter worked 13 years in fields of reactor physics and fuel materials science for advance concepts at Knolls Atomic Power Laboratory.

In 2006, Jonathan left to work for AREVA (now back to Framatome) in Lynchburg, VA for the U.S. version of the Evolutionary Power Reactor (USEPR); a commercial nuclear power plant design certification application process, where he branched out the fuel performance, plant systems safety analysis, and reactor I&C, culminating with development of a new methodology for control rod ejection safety analysis.

Before transitioning to NTP project with BWXT, Dr. Witter's tenure at BWXT began with the engineering efforts of the Small Modular Reactor mPower project, where he lead systems integration and design analysis in areas that had the tightest interface of operational performance and I&C controls for safety and monitoring.

A3.6 UNIVERSITY

Fig. A3.23 Dr. Dale Thomas

Dr. L. Dale Thomas joined the University of Alabama Huntsville (UAH) in 2015 as a full professor and was Board-appointed as the Eminent Scholar in Systems Engineering. He established the UAH Complex Systems Integration Laboratory (CSIL); an advanced systems engineering research facility focusing on Model-Based Systems Engineering.

In July 2017, he was appointed as the Deputy Director of the UAH Propulsion Research Center (PRC). In that role, he leads the Propulsion Systems Engineering research team and engages in strategic planning activities. Among several projects, the lab is assisting the NASA Marshall Space Flight Center (MSFC) with cube satellites deployed during the first Space Launch System (SLS) in 2019.

Prior to joining the UAH, Dr. Thomas worked for the NASA Marshall Space Flight Center. He began his NASA career in 1983 as an aerospace engineer in Marshall's Systems Analysis & Integration Laboratory. He was a project engineer from 1988 to 1993, managing selected disciplines in the detailed design and development activity for the International Space Station. He held the following positions over the years:

- Associate Director, Office of the Center Director, 2011
- Deputy Program Manager, Constellation Program, 2007-2010
- Chief of the Systems Engineering, Spacecraft & Vehicle Systems Department, 2006-2007
- Manager, Systems Engineering and Integration Office, 2004- 2006
- Director of the Systems Management Office, 2002 -2004
- Manager of Marshall's Systems Engineering and Integration 2001-2002
- Manager of the Marshall Center's Systems Engineering Office, 1999 to 2001
- Chief of the Systems Test Division from 1998 to 1999,
- Technical Assistant to the Director of the Systems Engineering & Integration Laboratory, 1996-1998
- Project Engineer at the Marshall Center from 1988 to 1993

Dr. Thomas received a B.S. in Industrial and Systems Engineering in 1981 from the UAH and a M.S in Industrial Engineering in 1983 from North Carolina State University. He earned a Ph.D. in Systems Engineering in 1988 from the UAH.

He has received numerous awards including
Presidential Rank of Meritorious Executive
NASA Outstanding Leadership Medal
Silver Snoopy Award

NASA Exceptional Service Medal and many others.

IMAGE LINKS

Fig. A3.1 Dr. Jacob Bleacher
https://www.nasa.gov/sites/default/files/styles/side_image/public/thumbnails/image/20190226-_1ag4182.jpg?itok=JddpWjo6

Fig. A3.2 Dr. Anthony Calomino
https://www.nasa.gov/sites/default/files/styles/side_image/public/thumbnails/image/calomino_anthony-for-bio.png?itok=RdnC9tmX

Fig. A3.3 Dr. Pam Meloy
https://www.nasa.gov/sites/default/files/styles/side_image/public/thumbnails/image/nhq202106250001_orig.jpg?itok=aFvWKINR

Fig. A3.4 James Reuter
https://www.nasa.gov/sites/default/files/styles/side_image/public/thumbnails/image/jim_reuter_0.jpg?itok=JTdhhhHW

Fig. A3.5 Dr. Stanley Borowski
https://www.aiaa.org/images/default-source/forums/speakers/stan-borowski.png?sfvrsn=591ea289_0

Fig. A3.6 Dr. William T. Emrich
https://www.facebook.com/photo/?fbid=10200494912334481&set=a.1343706451876&__tn__=%3C

Fig. A3.7 Dr. Daniel Herman
https://lh3.googleusercontent.com/yxy4x0LTP0ULo7qTsTVRuyaDDaggvlX0InmHRq3jEIoVo-iD2GvYCrzeHOf9tCQ9wkC4WGYOqKvXhIl9BdKuq2z5pFTwy_f9Z91bOGu0

Fig. A3.8 Dr. Michael Houts

https://lh3.googleusercontent.com/fcFi0_fE-e15WkoAnwzD4o_Q06W-gci4DHA4cWhduB7XxjiW27T03VwvCwzYwoAMUTGzm-_0FJIST94sEGnfdmsxC0A-WZdiZVmGFErOig

Fig. A3.9 Sonny Mitchell
https://www.nasa.gov/sites/default/files/thumbnails/image/sonny_full-size10.jpg

Fig. A3.10 Michelle Rucker
https://images-cdn.dashdigital.com/acschemmatters/february_2021/data/articles/img/015-01.jpg

Fig. A3.11 Dr. George Williams GRC
https://lh3.googleusercontent.com/dAdGhujF3XKpuJw4YOYPD2-KcG9mkHOWuu7j2EYNRheebh4ypQ017-i4krVagW-RPkIBlsEKE0KAS9ABumhFTPciEAKOt0y4HsH8EK68

Fig. A3.12 Tabitha Dodson
https://lh3.googleusercontent.com/fJXGIEhC_7DFC3Xcmaq5kkq3ESBB1QpfLN-

Fix A3.13 Dr. Stefanie Tompkins
https://media.defense.gov/2021/Aug/02/2002819651/-1/-1/0/210802-D-D0439-030.JPG

Fig. A3.14 Dr. Stephen Johnson
https://bios.inl.gov/BioPhotos/Steve_Johnson.jpg

Fig. A3.15 Michael B. Smith
https://www.ornl.gov/sites/default/files/styles/staff_profile_image_style/public/2020-02/Smith%2C%20Michael.jpg?h=01c99a79&itok=Fgup60fe

Fig. A3.16 Dr. John Wagner

https://bios.inl.gov/BioPhotos/JohnWagner.jpg

Fig. A3.17 Michael Eades USNC
https://lh3.googleusercontent.com/SGb1ds-
U1J3PU2c2VlH42JiM3HCovActAUR6GxNNp1L2eZl3Z599s0byDbj3LJ86S
5VgVe3r69ZoiB_JKCrshmGqhOrqfDHzfVi-siiG

Fig. A3.18 David Durham
https://www.westinghousenuclear.com/Portals/0/about-
2020/leadership/David_Durham_color_140.jpg

Fig. A3.19 Dr. Pete Pappano
https://lh3.googleusercontent.com/VfCb235NGu6-
eGMeCxQP4tqrvitEabxTXdx_qa5k3xgjFldX6_kGK9q6UuEYafqqmk6rM2ty
wf6hMckQBTJWkG4ukchvACUlP8QS-odG

Fig. A3.20 Dr. Brad Rearden, X-energy
https://lh3.googleusercontent.com/Wa_2dXjLIDUr_8wtn_hA5mq3Hnyrbnphh
pGSJgPypVkGiVL7-zqA4XRe0AXs-
BN7GW3MAKLZHoMik3oA670yBtH7joc39B6RyF5OcJIc

Fig. A3.21 Dr. Paolo Venneri
https://lh3.googleusercontent.com/dsYtqmdRCNYrPzmdWhYmQwluXL53gf
HLt_qm3irRIoL2IYw6TqycgQo-
HWbcz4ojyUx7Kfg9p9X9f9s8SA1wlmqoZZWf159umjy1qdyP

Fig. A3.22 Dr. Jonathan Witter
https://lh3.googleusercontent.com/4vpUbfCrlRbda9CI1cuDgzMwIlM3jxbjShx
rqMsC9GF7LYonjhqYaF2vVSw4OHaxgQJHJOhcIHGMXghk6KaNyVIGFrh
yd4ecdA5aE2Q

Fig. A3.23 Dr. Dale Thomas
https://www.uah.edu/images/news/campus/drdalethomaswide.jpg

Appendix 4

Quotes

"Although the undertaking of this mission (a crewed Nuclear Thermal Propulsion rocket to Mars) will be a great national challenge, it represents no greater challenge than the commitment made in 1961 to land a man on the moon."
Wernher von Braun, Director of the Marshall Space Flight Center

"Secondly, an additional 23 million dollars, together with 7 million dollars already available, will accelerate development of the Rover nuclear rocket. This gives promise of someday providing a means for even more exciting and ambitious exploration of space, perhaps beyond the Moon, perhaps to the very end of the solar system itself."
Excerpt from the 'Special Message to the Congress on Urgent National Needs' President John F. Kennedy Delivered in person before a joint session of Congress May 25, 1961

"I have long felt that to truly open the solar system to human exploration some type of nuclear engine will be an absolute necessity."
"I just feel very fortunate to have had the privilege of being able to work with such a great group of people on such a wonderful and challenging project," he says. *"To work on advancing rocket engines like these has been my dream since I was a little kid. I can still hardly believe that I actually get to do it."*
Dr. William J Emrich, Manager, the NTREES facility at Marshall.

"NASA will work with our long-term partner, DARPA, to develop and demonstrate advanced nuclear thermal propulsion technology as soon as 2027. With the help of this new technology, astronauts could journey to and from deep space faster than ever; a major capability to prepare for crewed

missions to Mars. Congratulations to both NASA and DARPA on this exciting investment, as we ignite the future, together."
NASA Administrator Bill Nelson.

"Nuclear power has opened the solar system to exploration, allowing us to observe and understand dark, distant planetary bodies that would otherwise be unreachable. And we're just getting started. Future nuclear power and propulsion systems will help revolutionize our understanding of the solar system and beyond and play a crucial role in enabling long-term human missions to the Moon and Mars."
Dr. Thomas Zurbuchen, Associate Administrator for NASA's Science Mission Directorate.

"Expanding our partnership (with DARPA) to nuclear propulsion will help drive forward NASA's goal to send humans to Mars."
NASA Deputy Administrator Pam Melroy.

"Nuclear thermal propulsion could be the ticket to Mars. The results from this study will give us a better idea of whether that is the case by experimentally measuring key factors related to engine performance and lifetime."

"This is vital testing, helping us reduce risks and costs associated with advanced propulsion technologies and ensuring excellent performance and results as we progress toward further system development and testing,"
Michael Houts, NTP Manager at the Marshall.

"The information we gain using this test facility (NTREES) will permit engineers to design rugged, efficient fuel elements and nuclear propulsion systems. It's our hope that it will enable us to develop a reliable, cost-effective nuclear rocket engine in the not-too-distant future."
Bill Emrich, Manager of the NTREES facility at Marshall.

"DARPA and NASA have a long history of fruitful collaboration in advancing technologies for our respective goals, from the Saturn V rocket that

took humans to the Moon for the first time to robotic servicing and refueling of satellites. The space domain is critical to modern commerce, scientific discovery, and national security. The ability to accomplish leap-ahead advances in space technology through the DRACO nuclear thermal rocket program will be essential for more efficiently and quickly transporting material to the Moon and Eventually, people to Mars."

Dr. Stefanie Tompkins, Director, DARPA.

"With this collaboration, we will leverage our expertise gained from many previous space nuclear power and propulsion projects. Recent aerospace materials and engineering advancements are enabling a new era for space nuclear technology, and this flight demonstration will be a major achievement toward establishing a space transportation capability for an Earth-Moon economy."

Jim Reuter, Associate Administrator for NASA STMD.

"The space domain is essential to modern commerce, scientific discovery, and national defense. Moving larger payloads into farther locations in cislunar space – the volume of space between the Earth and the Moon – will require a leap-ahead in propulsion technology.

Dr. Tabitha Dodson, Chief of Communications, DARPA.

"You're not burning anything, so you don't have to carry oxygen, which is really heavy. But you still have to carry a lot of hydrogen all the way out to Mars so you can get back and that requires an extremely efficient system. NTP is tailor made for that."

Mike Kynard, Former Project Manager for NASA's Space Nuclear Propulsion Project.

"NASA and the DOE are collaborating on this important and challenging development that, once completed, will be an incredible step towards long-term human exploration of the Moon and Mars. We'll take advantage of the unique capabilities of the government and private industry to provide reliable, continuous power that is independent of the lunar location."

Todd Tofil, Fission Surface Power Project Manager at NASA's Glenn Research Center.

"I'm a real advocate for nuclear thermal propulsion because you need to build out that Infrastructure for higher power levels that solar or battery systems just can't provide. We want to help NASA ensure that nuclear thermal propulsion is a viable option and to bring it into the next level of technology maturation."
Dr. Witter of BWX Technologies

"NTP is an extraordinarily exciting new program for BWXT. The exploration of space is a noble endeavor, and no one is better qualified to design and build reactors for the remote and challenging environment of space than BWXT."
 Rex Geveden, BWXT President and CEO

"Space nuclear power and propulsion is a game-changing concept that could unlock future deep-space missions that take us to Mars and beyond. This study will help us understand the exciting potential of atomic-powered spacecraft, and whether this nascent technology could help us travel further and faster through space than ever before."
Dr. Graham Turnock, Chief Executive of the UK Space Agency

"We are excited to be working with the UK Space Agency on this pioneering project to define future nuclear power technologies for space. We believe there is a real niche UK capability in this area and this initiative can build on the strong UK nuclear network and supply chain. We look forward to developing this and other exciting space projects in the future as we continue to develop the power to protect our planet, secure our world and explore our universe"
Dave Gordon, UK Senior Vice President, Rolls-Royce Defence

"The longer you're out there, the more time there is for stuff to go wrong".
Jeff Sheehy, Chief Engineer of NASA's Space Technology Mission Directorate

Appendix 5

Nuclear Power in Space

A5.1 Background

The United Nations has an Office for Outer Space Affairs (UNOOSA) which implements the decisions of the Committee on the Peaceful Uses of Outer Space (COPUOS) set up in 1959 and now with 71 member states. UNOOSA recognizes "that for some missions in outer space nuclear power sources are particularly suited or even essential owing to their compactness, long life and other attributes" and "that the use of nuclear power sources in outer space should focus on those applications which take advantage of the particular properties of nuclear power sources." It has adopted a set of principles applicable "to nuclear power sources in outer space devoted to the generation of electric power on board space objects for non-propulsive purposes," including both radioisotope systems and fission reactors.

In the early days of the space program, chemical propulsion was the obvious way to go to space; it was, after all, proven in WWII and during the Cold War. Today, the use of nuclear power for propulsion purposes has come to the forefront for deep space propulsion; but not for the launch phase.

"Nuclear power has opened the solar system to exploration, allowing us to observe and understand dark, distant planetary bodies that would otherwise be unreachable. And we're just getting started," said Dr. Thomas Zurbuchen, associate administrator for NASA's Science Mission Directorate. "Future nuclear power and propulsion systems will help revolutionize our understanding of the solar system and beyond and play a crucial role in enabling long-term human missions to the Moon and Mars."

Six decades after the launch of the first nuclear-powered space mission, Transit IV-A, NASA is embarking on a bold future of human exploration and scientific discovery. This future builds on a proud history of safely launching and operating nuclear-powered missions in space.

On June 29, 1961, the Johns Hopkins University Applied Physics Laboratory launched the Transit 4A (also Transit IV-A) spacecraft. It was a U.S. Navy navigational satellite with a SNAP-3B radioisotope powered generator producing 2.7 watts of electrical power; about enough to light an LED bulb. (Even a night light is 4 watts).Transit IV-A broke an APL mission-duration record and confirmed the Earth's equator is elliptical. It also set the stage for ground-breaking missions that have extended humanity's reach across the solar system.

Fig. A5.1 Transit 4A. Photo courtesy of Johns Hopkins University Applied Physics Laboratory.

The sphere at the top is Greb III, designed to measure solar X-rays. In the middle is Injun; an experiment to record the flux of charged particles responsible for the aurora and airglow. The Transit satellite is the large sphere below.

Since 1961, NASA has flown more than 25 missions carrying a nuclear power system through a successful partnership with the Department of Energy which provides the power systems and plutonium-238 fuel. "The department and our national laboratory partners are honored to play a role in powering NASA's space exploration activities," said Tracey Bishop, deputy assistant secretary in DOE's Office of Nuclear Energy. "Radioisotope Power Systems are a natural extension of our core mission to create technological solutions that meet the complex energy needs of space research, exploration, and innovation." They are a category of power systems that convert heat generated by the decay of plutonium-238 fuel into electricity.

A5.2 Radioisotope Thermoelectric Generators (RTGs)

There are only two practical ways to provide long-term electrical power in space: the light of the Sun or heat from a nuclear source. "As missions move farther away from the Sun to dark, dusty, and harsh environments, like Jupiter, Pluto, and Titan, they become impossible or extremely limited without nuclear power," said Leonard Dudzinski, Chief Technologist for NASA's Planetary Science Division and program executive for Radioisotope Power.

The Multi-Mission Radioisotope Thermoelectric Generator (MMRTG) for short or just RTGs, have been the main power source for the U.S. space work since 1961. The high decay heat of Plutonium-238 (0.56 W/g) enables its use as an electricity source in the RTGs of spacecraft, satellites and navigation beacons. Its intense alpha decay process with negligible gamma radiation calls for minimal shielding. Americium-241, with 0.15 W/g, is another source of energy, favored by the European Space Agency, though it has high levels of relatively low-energy gamma radiation. Heat from the oxide fuel is converted to electricity through static thermoelectric elements (solid-state thermocouples), with no moving parts. RTGs are safe, reliable and maintenance-free and can provide heat or electricity for decades under very harsh conditions, particularly where solar power is not feasible.

The importance of such power sources was illustrated by the European Space Agency's Rosetta mission, which successfully landed the Philae probe on comet 67P/Churymov–Gerasimenko in 2014. Equipped with batteries and solar panels, the position in which Philae came to rest on the comet's surface was shielded from the Sun's rays by cliffs, meant that the lander was unable to make use of solar energy and was only able to send 64 hours' worth of data before its battery power ran out.

So far, over 45 RTGs have powered in excess of 25 U.S, space vehicles including Apollo, Pioneer, Viking, Voyager, Galileo, Ulysses, Cassini and New Horizons space missions, as well as many civil and military satellites. The probe's lifetime is presently limited only by its nuclear fuel supply, which is likely sufficient to keep New Horizons flying through 2040; and NASA recently granted another mission extension for New Horizons. "The RTG was, and still is crucial to New Horizons," said Alan Stern, New Horizons Principal Investigator from the Southwest Research Institute (SwRI). "We couldn't do the mission without it. No other technology exists to power a mission this far away from the Sun, even today."

The Cassini spacecraft carried three RTGs providing 870 watts of power from 33 kg (74 lbs) of plutonium-238 oxide as it explored Saturn. It was launched in 1997, entered Saturn's orbit in 2004, and functioned very well until it was terminated in September 2017.

Voyager 1&2 spacecraft which have sent back pictures of distant planets have already operated for over 35 years since 1977 launch and are expected to send back signals powered by their RTGs through to 2025. Galileo, launched in 1989, carried a 570-watt RTG. The Viking and Rover landers on Mars in 1975 depended on RTG power sources, as does the Mars Science Laboratory rover Curiosity launched in 2011.

Three RTGs (each with 2.7 grams of plutonium-238 dioxide) were used as heat sources on the Pathfinder Mars robot lander launched in 1996, producing 35 watts. Each produced about one watt of heat. The 10.5 kg (23 lbs) Pathfinder rovers in 1997 and the two Mars rovers operating during 2004-09, used solar panels and batteries, with limited power and life. The latest plutonium-powered RTG is a 290-watt system known as the GPHS RTG. The thermal power for this system is from 18 general purpose heat source (GPHS) units. Each GPHS contains four iridium-clad ceramic Pu-238 fuel pellets, stands 5 cm tall, 10 cm square and weighs 1.44 kg. The multi-mission RTG (MMRTG) (see image below) uses eight GPHS units with a total of 4.8 kg of plutonium oxide producing 2 kW thermal which can be used to generate some 110 watts of electric power, 2.7 kWh/day.

Today, a MMRTG powers the Perseverance rover, which is captivating the nation as it searches for signs of ancient life on Mars The Perseverance rover's RTG carries 4.8 kg (10.7 lbs) of Pu-238 and reliably produces about 110 watts (similar to a light bulb) at the start of the mission and declines over time.

Flight Systems for Recent Missions

Multi-Hundred Watt –
Radioisotope Thermoelectric
Generator (MHW–RTG)
(Voyager)

General Purpose Heat Source –
Radioisotope Thermoelectric
Generator (GPHS–RTG)
(Galileo, Ulysses, Cassini)

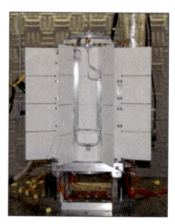

Current RPS:
Multi-Mission Radioisotope
Thermoelectric Generator
(MMRTG)
(Curiosity,
Mars 2020,
Dragonfly)

Fig. A5.2 Examples of RTG's. Photo courtesy of JPL/NASA.

A5.3 Radioisotope Power Systems

Radioisotope power systems (RPS) are a type of nuclear energy technology that uses heat to produce electric power for operating spacecraft systems and science instruments. That heat is produced by the natural radioactive decay of plutonium-238. Choosing between solar and nuclear power for a space mission has everything to do with where a spacecraft needs to operate and what the mission must accomplish when it gets there. Radioisotope power is used only when it will enable or significantly enhance the ability of a mission to meet its science goals.

RPS offer several important benefits. They are compact, rugged and provide reliable power in harsh environments where solar arrays are not practical. For example, Saturn is about ten times farther from the sun than Earth, and the available sunlight there is only one hundredth, or one percent, of what we receive at Earth. At Pluto, the available sunlight is only six hundredths of a percent of the amount available at Earth. The ability to utilize radioisotope power is important for missions to these and other incredibly distant destinations, as the size of solar arrays required at such distances is impractically large with current technology.

RPS offer the key advantage of operating continuously over long-duration space missions, largely independent of changes in sunlight, temperature, charged particle radiation, or surface conditions like thick clouds of dust.

In addition, some of the excess heat produced by some radioisotope power systems can be used to enable spacecraft instruments and on-board systems to continue to operate effectively in extremely cold environments.

RPS are sometimes referred to as a type of "nuclear battery." While some spacecraft, like Cassini, did run their systems directly off of their RPS, others like the Mars Science Laboratory rover can use the RPS to charge batteries and run their systems and instruments off of stored battery power. In either case, the RPS is attached directly to a spacecraft, much like a power cord being plugged in.

These technologies are capable of producing electricity and heat for decades under the harsh conditions of deep space without refueling. All of these power systems, flown on more than two dozen NASA missions since the 1960s, have functioned for longer than they were originally designed.

The RPS used to power NASA spacecraft are supplied by the DOE. NASA and DOE continue to collaborate on maintaining and developing several types of RPS.

A5.3.1 GPHS

The General Purpose Heat Source module, or GPHS, is the essential building block for the radioisotope generators used by NASA. These modules contain and protect the plutonium-238 (or Pu-238) fuel that gives off heat for producing electricity. The fuel is fabricated into ceramic pellets of plutonium-238 dioxide and encapsulated in a protective casing of iridium, forming a fueled clad. Fueled clads are encased within nested layers of carbon-based material and placed within an aeroshell housing to comprise the complete GPHS module.

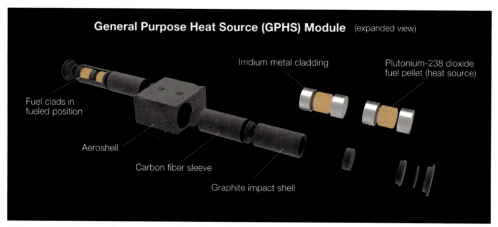

Fig. A5.3 Expanded view of a GPHS. Illustration courtesy of NASA,

For a 0:36 sec video showing the assembly of a GPHS, go to: https://youtu.be/dHwFxK-6fkE

Each GPHS is a block about four by four by two inches in size, weighing approximately 1.5 kg (3.5 lbs). They are nominally designed to produce thermal power at 250 watts at the beginning of a mission, and can be used individually or stacked together.

Modules have been subjected to extreme testing conditions that significantly exceeded the intensity of a wide range of potential accidents. Such tests have included simulating multiple reentries for a single module through Earth's atmosphere, exposure to high temperature rocket propellant fires, and impacts onto solid ground.

The enhanced GPHS modules used in the latest generation of radioisotope power systems incorporate additional rugged, safety-tested features that build upon those used in earlier generations. For example, additional material (20 percent greater in thickness) has been added to the graphite aeroshell and to the two largest faces of the block-like module. These modifications provide even more protection to help to contain the fuel in a wide range of accident conditions, further reducing the potential for release of plutonium-238 that might result.

A5.3.2 RHU's

Radioisotope Heater Units (RHUs) have been called the "unsung heroes" of radioisotope power technology. Most spacecraft can use solar energy to provide heat for keeping their structures, systems, and instruments warm enough to operate effectively. But when other heat source technologies are not feasible, an alternate heat source is required for the spacecraft.

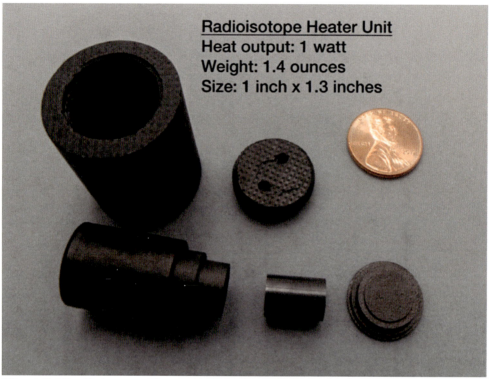

Fig. A5.4 Radioisotope Heater Unite. Illustration courtesy of DOE.

For a 38 sec animation of a RHU, go to:
https://youtu.be/fHH_yGtmKZs

RHUs are small devices that use the decay of plutonium-238 to provide heat to keep spacecraft components and systems warm so that the equipment can survive long enough in the cold space environment to complete its mission. This heat is transferred to spacecraft structures, systems, and instruments directly, without moving parts or intervening electronic components.

By using RHUs, the spacecraft designer can allocate scarce spacecraft electrical power to operate the spacecraft systems and instruments. RHUs also provide the added benefit of reducing potential interference (electromagnetic interference) with instruments or electronics that might be generated by electrical heating systems.

Radioisotope Heater Units have been critical for providing heat to keep some spacecraft warm enough to accomplish their missions, including the battery powered Galileo and Huygens probes and the two solar powered Mars Exploration Rovers.

An RHU contains a Pu-238 fuel pellet about the size of a pencil eraser and outputs about 1 Watt of heat. (The entire RHU is about the size of a C-cell battery.) Some missions employ just a few RHUs for extra heat, while others have dozens.

NASA missions enabled by radioisotope heater units

Apollo 11 EASEP Lunar Radioisotope Heater - contained two 15W RHUs

Pioneer 10 & 11 - 12 RHUs each

Voyager 1 & 2 - 9 RHUs each

Galileo - 120 RHUs (103 on orbiter, 17 on atmospheric probe)

Mars Pathfinder Sojourner Rover - 3 RHUs

Cassini - 117 RHUs (82 on orbiter, 35 on Huygens Titan probe)

MER Spirit & Opportunity Rovers - 8 RHUs each

A5.4 Future Missions

Dragonfly, which is set to launch in 2027, is the next mission with plans to use an MMRTG. Part of NASA's New Frontiers program, Dragonfly is an octocopter designed to explore and collect samples on Saturn's largest moon, Titan, an ocean world with a dense, hazy atmosphere.

Dragonfly will use a large lithium-ion battery as its power source, which is designed to be recharged by a MMRTG between flights. The battery itself uses the same cells that are currently used on the International Space Station, and the MMRTG is the same kind as on the Curiosity and Perseverance rovers on Mars. Both the battery and MMRTG power source will be enclosed in an insulated cylinder at the rear of the lander to protect them from the from the minus 290° F Titan atmosphere. The MMRTG also produces heat, which can be used to keep the interior of Dragonfly warm.

Fig. A5.5 The RTG powered Dragonfly to Titan. Photo courtesy of NASA.

"Radioisotope Power Systems (RPS) is really an enabling technology," said APL's Zibi Turtle, Principal Investigator for the upcoming Dragonfly mission. "Early missions like Voyager, Galileo, and Cassini that relied on RPS have completely changed our understanding and given us a geography of the distant solar system. Cassini gave us our first close-up look at the surface of Titan." According to Turtle, the MMRTG serves two purposes on Dragonfly: power output to charge the lander's battery and waste heat to keep its instruments and electronics warm.

"Flight is a very high-power activity. We'll use a battery for flight and science activities and recharge the battery using the MMRTG," said Turtle. "The waste heat from the power system is a key aspect of our thermal design. The surface of Titan is very cold, but we can keep the interior of the lander warm and cozy using the heat from the MMRTG."

As the scientific community continues to benefit from RPS, NASA's Space Technology Mission Directorate is investing in new technology using reactors and low-enriched uranium fuel to enable a robust human presence on the Moon and eventually human missions to Mars.

Astronauts will need plentiful and continuous power to survive the long lunar nights and explore the dark craters on the Moon's South Pole. A fission surface power system could provide enough juice to power robust operations. NASA is leading an effort, working with the DOE and industry, to design a fission power system for a future lunar demonstration that will pave the way for base camps on the Moon and Mars. See Chapter 6.6.

IMAGE LINKS

Fig. A5.1 Transit IV
https://rps.nasa.gov/rails/active_storage/blobs/eyJfcmFpbHMiOnsibWVzc2Fn ZSI6IkJBaHBBbnNGIiwiZXhwIjpudWxsLCJwdXIiOiJibG9iX2lkIn19-- dbd7f5cae7c3875ce56ad53bae6e455960904131/Transit-4a_and_Injun- 1_and_SolRad-3_image2.jpg?disposition=attachment

Fig. A5.2 Flight Systems for Recent Missions
https://slideplayer.com/slide/17496502/102/images/4/Flight+Systems+for+Rec ent+Missions.jpg

Fig. A5.3 GPHS
https://rps.nasa.gov/rails/active_storage/blobs/eyJfcmFpbHMiOnsibWVzc2Fn ZSI6IkJBaHBBDZz09IiwiZXhwIjpudWxsLCJwdXIiOiJibG9iX2lkIn19-- fca66eb63d80dcfdb053dd760bf7c60448688899/GPHSStill_20130627_labeled .jpg?disposition=attachment

Fig. A5.4 RHU
https://rps.nasa.gov/rails/active_storage/blobs/eyJfcmFpbHMiOnsibWVzc2Fn ZSI6IkJBaHBBBYlk9IiwiZXhwIjpudWxsLCJwdXIiOiJibG9iX2lkIn19-- 7858ae950a06d22df2153e18b8adc31374f963a7/RHUnew_labeled_7201.jpg?d isposition=attachment

Fig. A5.5 Dragonfly
https://astrobiology.nasa.gov/uploads/filer_public_thumbnails/filer_public/cf/2 3/cf235c8f-7b21-4ba7-bf34-42db0785cb1e/hero_dragonfly- 01.jpg__1240x510_q85_crop_subject_location-620%2C254_subsampling- 2.jpg

Appendix 6

Russian Nuclear Rocket Programs

A6.1 History of Russian Space Reactors

Thanks to Astronautix.com for the following article which follows:
"Soviet nuclear thermal propulsion was initially pursued as an alternative to nuclear electric propulsion. The earliest work was a follow-on to OKB-1's late 1950's designs for missiles and launch vehicles using nuclear thermal engines with ammonia as propellant. OKB is a transliteration of the Russian initials meaning "experiment and design bureau". During the Soviet era, OKBs were closed institutions working on design and prototyping of advanced technology, usually for military applications. The English language corresponding term for such bureau's occupation is: Research and Development.

This effort had been abandoned in 1960 due to the long development scale and the safety problems involved in testing nuclear systems in the first stage of rockets. Thereafter, the engine bureaus of Bondaryuk (OKB-670) and Glushko (OKB-456) continued study of nuclear propulsion, but using liquid hydrogen for upper stage applications. Engines of 200 metric tons (220 US tons) and 40 metric tons (44 US tons) thrust with a specific impulse of 900 to 950 seconds were being considered. At the end of 1961 both bureaus completed their draft projects and it was decided to continue work on development of an engine in the 30 to 40 metric ton (33-40 US ton) thrust range. In the following year, Sergei Korolev was asked to study application of such engines, followed by a demand in May, 1963 from the Scientific-Technical Soviet for specific recommendations.

Korolev considered three variants based on the N1:

1. A three stage vehicle using the N1 first and second stages and a nuclear third stage,

2. A three stage vehicle using the N1 first stage and nuclear second and third stages,

3. A two stage vehicle using the N1 first stage and a nuclear second stage

Considered for each case were nuclear engine designs Type A (18 metric tons thrust, 4.8 metric tons mass), AF (20 metric tons thrust, 3.25 metric tons mass), V (40 metric tons thrust, 18 metric tons mass), and V with a bioshield for use on manned flights (40 metric tons thrust, 25 metric tons mass).

The study concluded that the two stage vehicle was the most promising. Compared to an equivalent vehicle using liquid oxygen/liquid hydrogen, mass in low earth orbit would be more than doubled. Optimal stage size was 700 to 800 metric tons (770-880 US tons) for the Type A engines and 1,500 to 2,000 metric tons (1650-2200 US tons) for the type V engines (this resulted in an extremely large number of nuclear engines by Western standards). Use of the nuclear stage would result in a single N1 launch being able to launch a round-trip lunar landing (mass landed on lunar surface over 24 metric tons (26 US tons) with return of a 5 metric ton (5.5 US tons) capsule to earth).

For a Mars expedition, it was calculated that the AF engine would deliver 40% more payload than a chemical stage, and the V would deliver 50% more. But Korolev's study also effectively killed the program by noting that his favored solution, a nuclear electric ion engine, would deliver 70% more payload than the LOX/LH2 stage.

Further investigation of nuclear thermal stages for the N1 does not seem to be pursued by OKB-1. Bondaryuk and Glushko turned to Chelomei and his MK-700 Mars spacecraft for future application of such stages. Glushko had designed his RD-410, 7 metric ton (7.7 US tons) thrust engine in the 1960's. But he also undertook an even more ambitious engine in the 1963 to 1970 period - the RD-600 gas core nuclear reactor. This exotic technology, also pursued in the United States, would have resulted in a 200 metric ton (220 US tons) thrust engine with a specific impulse of 2000 seconds. Although the draft project was completed and it was concluded the concept was entirely feasible, no funds for development were forthcoming.

It was not Glushko or Bondaryuk, but the Kosberg OKB that took concrete steps in the 1960's toward actually building nuclear thermal propulsion system hardware. In 1962, Kosberg, together with the Kurchatov Institute, Keldysh Scientific Centre, NPO Luch, and a half dozen other institutes, began experiments with nuclear thermal rockets for upper stage applications that used liquid hydrogen as the propellant.

However, a 1972 review by a government expert commission of Chelomei's MK-700 Mars spacecraft design, dealt a mortal blow to further rapid development of nuclear thermal propulsion. Among other findings, the commission, noting the American abandonment of manned Mars plans and their NERVA engine, found no pressing need for an equivalent Soviet project. They noted that Russian nuclear thermal engines were only in the draft project stage and would take 15 to 20 years to reach technological maturity. It was felt that the radiation safety problem of nuclear thermal propulsion had only been solved theoretically. Negotiations with the United States would be required to achieve international permission for placing large nuclear reactors into orbit. The outcome of these negotiations was uncertain.

The state commission recommended that further work on manned Mars expeditions be deferred indefinitely. However Soviet development of nuclear-thermal propulsion was allowed to continue. Although Glushko abandoned the technology, NPO Luch began tests of prototype Kosberg engines at a test stand 50 km (31 mi) southwest of Semipalatinsk-21 in 1971. Tests continued there through 1978. Simultaneously, a more elaborate facility was built 65 km (40 mi) south of Semipalatinsk-21 for comprehensive tests of the Baikal-1 prototype engine. Thirty simulated flights were conducted from 1970 to 1988 without failure. It was eventually proposed that two engines would be derived from this work: the RD-0410, a "minimum" engine, of 3.5 metric tons (38.6 US tons) thrust; and later the RD-0411, a 70 metric ton (77 US tons) thrust engine.

Fig. 6.2 RD-0410 Thermal Nuclear Rocket. Photo Credit: © Dietrich Haeseler

This rocket was developed by the Chemical Automatics Design Bureau in Voronezh from 1965 through the 1980s using liquid hydrogen propellant. The engine was ground-tested at the Semipalatinsk Test Site, and its use was incorporated in the Kurchatov Mars 1994 crewed mission proposal.

This engine had slightly higher performance (exhaust temperature and specific impulse) over NERVA; the U.S. nuclear thermal rocket engine project. The design of the reactor core included thermal insulation between uranium carbide/tungsten carbide fuel and the zirconium hydride moderator. This allowed for a very compact reactor core design. Hydrogen flow cooled the moderator. First, it allowed keeping a very low neutron energy and high fission cross-section; then it was heated by the direct contact to the fuel rods. To prevent the chemical reaction between carbide and hydrogen, about 1 percent of heptane was added to the hydrogen after the moderator passage.

By 1989, work on both nuclear electric and nuclear thermal propulsion included bimodal use of the nuclear reactors to provide electrical power during dormant or ballistic cruise phases of flight. In the case of nuclear thermal engines, this meant addition of a Brayton cycle turbine using xenon-helium coolant. A nuclear thermal Mars spacecraft proposed by the Kurchatov Institute in 1989 featured a new design power plant of 20,000 kgf (44,092 lbf), a thermal power of 1200 MW, operating time of 5 hours, and a specific impulse of between 815 and 927 seconds. During cruise operations the turbine would provide 50-200 kW of electric power, requiring 600 m² (6,458 ft²) of radiators at the end of the spacecraft. Total mass of this combination power plant was estimated to be 50 to 70 metric tons (55-77 US tons).

The collapse of the Soviet Union ended further development work on nuclear thermal propulsion."

A6.2 Current Russian Efforts

The following is an edited article by well know Russian journalist, Anatoly Zak published in RussianSpaceWeb.com on December 3, 2022. Edits include the deletion of Russian acronyms, referrals to animations and other edits that are compatible with this book. The following begins the quoted article:

"Russia reveals a formidable nuclear-powered space tug. After years of near silence, a prominent developer of Russian military spacecraft suddenly publicly floated the first pictures of a massive nuclear-powered space transport undergoing assembly at the company's facility in St. Petersburg. The KB Arsenal design bureau, which serves as the prime contractor in the project, is known for its Soviet-era nuclear-powered satellites, one of which infamously crashed in the Arctic region of Canada in 1977.

What is the nuclear-powered space tug?
A series of photos and computer-generated imagery, which surfaced on the Internet in 2020 and originated from KB Arsenal, revealed the apparent latest version and the planned operation of a very large space tug propelled by electric engines and powered by a nuclear source.

The project officially known as the Transport and Energy Module, (TEM), has been well known to the watchers of the Russian space program for more than a decade. Tracing its roots to the dawn of the Space Age, the TEM concept is attempting to marry a nuclear reactor with an electric rocket engine. The electric propulsion systems heat up and accelerate ionized gas to create a thrust-generating jet and, therefore, are alternatively known as ion or plasma engines. When measured per unit of spent propellant mass, electric engines are more efficient than traditional liquid or solid-propellant rockets, but their thrust is relatively low at any given time and they require a great deal of electric power to operate. Because of this, until recently, the practical use of electric propulsion in space flight was mostly limited to orbit adjustment systems aboard satellites or to deep-space missions, in which spacecraft could take advantage of low thrust over very long periods of time.

To scale up the operation of power-hungry electric thrusters, engineers long considered replacing heavy and bulky solar panels with nuclear power sources which could provide plenty of electricity for years, if not decades, and would not be dependent on solar radiation in the remote and cold regions of the Solar System, as demonstrated by planetary missions such as Voyager, Cassini and many others. However, the development of nuclear reactors for space still had to take place on Earth, where environmental and safety concerns slowed down the progress in this field.

Still, by the early 21st century, the Russian military apparently renewed interest in the great capacity of nuclear reactors to provide electricity not only for propulsion systems, but also for other equipment aboard large spacecraft, such as powerful radar antennas for surveillance purposes or anti-satellite lasers capable of blinding sensors aboard enemy spacecraft.

With a large portfolio of nuclear technology and a hefty budget, the Russian Ministry of Defense seemingly became the primary backer of the first post-Soviet attempt to build a nuclear power-generating system for space. Not surprisingly, the work on the reactor was largely classified, but in 2020, KB Arsenal released photos showing what appeared to be the assembly of the full-scale TEM vehicle or its prototype deployed in orbit.

Fig. A6.2 TEM in orbit. Photo courtesy of the KB Arsenal/Michael Jerdev.

In its corporate brochure, KB Arsenal reported that between 2016 and 2018, the company had conducted several early studies and preliminary designs looking into a megawatt-class nuclear-powered spacecraft. The program included the development and testing of the Design and Technical Mockups of the TEM module and its components, such as the Truss Carrier Section, the Support Systems Block, the Propulsion Unit Module and the Power Unit. These mockup elements have undergone functional tests, according to KB Arsenal. The company also published new photos of the spacecraft elements during assembly and testing.

In 2018 and 2019, Arsenal conducted the core study, which looked at military and civilian applications of the megawatt-class power module, including its implied use as an anti-satellite weapon. The potential tasks under consideration, as defined by Arsenal, included remote-sensing of the Earth's surface and of the airspace, electromagnetic impact on the radio-electronic command and control assets, reconnaissance, communications, navigation, inter-orbital transport and cargo delivery to near-lunar orbits. The proposals also included using the module for powering data-relay spacecraft in the

Martian L1 Lagrange point to provide communications between a base on the surface of Mars, Mars orbiters and the Earth. The delivery of power-producing nuclear systems to the Martian surface base was also mulled.

The heart of the TEM tug is a nuclear reactor, which generates heat. The heat is then converted into electrical power either through a mechanical turbine or via the so-called thermal emission method, which does not involve any moving parts. Though less effective than a turbine, the simpler, and more familiar to the Russian industry, thermo-emission conversion appeared to be in use aboard the TEM vehicle revealed in 2020.

The excessive heat energy inevitably generated in the process of reactor work is released into space with a system of radiators, which can also use a variety of different technologies to operate in weightlessness and beyond the atmosphere. The revealed TEM vehicle appeared to feature a trio of main and three auxiliary radiators. The latter smaller panels probably service the traditional needs of service systems aboard the spacecraft, while the larger deployable and stationary radiators were probably exclusively designed for removing the reactor's heat. The animation showed a very complex three-stage process of the main radiator deployment aboard the TEM module.

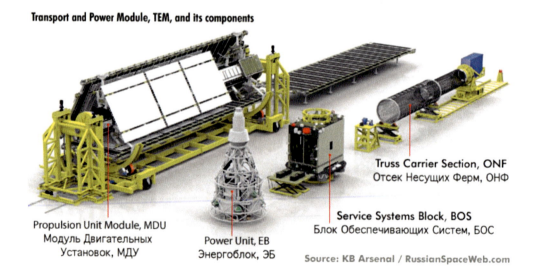

Transport and Power Module, TEM, and its components

Propulsion Unit Module, MDU
Модуль Двигательных
Установок, МДУ

Power Unit, EB
Энергоблок, ЭБ

Service Systems Block, BOS
Блок Обеспечивающих Систем, БОС

Truss Carrier Section, ONF
Отсек Несущих Ферм, ОНФ

Source: KB Arsenal / RussianSpaceWeb.com

Fig. A6.3 TEM components. Illustration courtesy of the KB Arsenal.

However, on the unveiled vehicle, the radiator panels appeared to be using heat-carrying cooling fluid pumped through the system by a turbine. It is a less progressive technology than the capillary heat pipe radiating system which was originally planned for the spacecraft and which Russia was known to be testing aboard the Mir space station at the turn of the 21st century.

To protect all the systems aboard the spacecraft from harmful radiation, the reactor is placed behind a cone-shaped shield which forms a protected conical "shadow" free of dangerous particles. To further increase the safe zone, the reactor is attached to what appeared to be a four-section telescopic boom made of a light-weight composite material. The boom deploys to its full length after the ship's separation from the launch vehicle in orbit.

According to the available publications, the nuclear reactor on the TEM vehicle would be activated only after the spacecraft reached a 600 or 800-km (373-497 mi) orbit, which is far enough from the rarified atmosphere to prevent the natural decay and reentry of a stalled satellite. In the interim, all the service systems of the space tug and its payloads could still receive power from a pair of solar panels deployed on the sides of the propulsion module immediately upon entering orbit.

The photos released by KB Arsenal in 2020, but likely showing a full-scale mockup assembled as early as 2018, showed key components of this very large vehicle, including the propulsion module, stationary and deployable radiators and the deployable boom which would carry the reactor. There were no photos of the reactor itself.

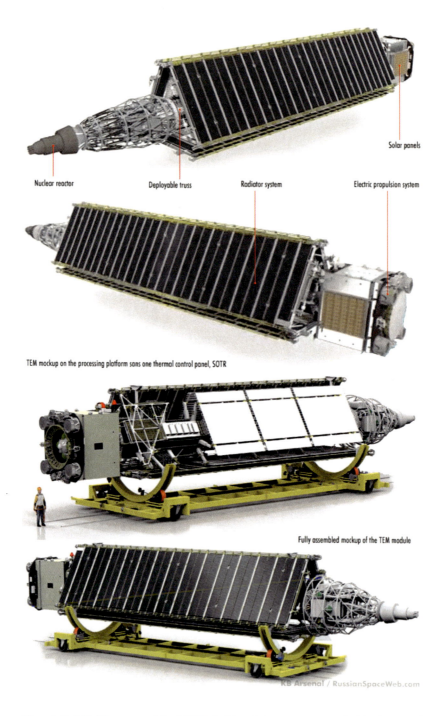

Solar panels

Nuclear reactor   Deployable truss   Radiator system   Electric propulsion system

TEM mockup on the processing platform sans one thermal control panel, SOTR

Fully assembled mockup of the TEM module

KB Arsenal / RussianSpaceWeb.com

Fig. A6.4 Fully assembled TEM. Graphic courtesy of KB Arsenal.

It appeared that even without its payload, the Russian TEM would be a 20 or 30-ton vehicle, which could require either Angara-5M or Angara-5V heavy rockets to enter an initial orbit from the Vostochny spaceport. One depiction produced by GKNPTs Khrunichev around 2016 showed the Angara-5V rocket with a Briz upper stage carrying the TEM vehicle.

During Moscow Air and Space Show, MAKS, opening on July 20, 2021, at Zhukovsky airfield near Moscow, Roskosmos displayed a scaled model of the nuclear-electric space tug for the Zevs (Zeus) complex. The exhibit appeared to include the initial experimental version of the vehicle equipped with ion engines and another scaled-up variant sporting the so-called rotor magnetic plasma engines.

The TEM vehicles were displayed alongside historic reactor-carrying US-A spacecraft which were developed during the Soviet period for guiding cruise missiles to their targets and appeared to be shown to the same scale with Zevs models, giving a general idea about the ambition of the current effort.

It is known that the operational version of the Zevs space tug was sized for launch on the Angara-5V rocket capable of delivering up to 38 tons of payload to the low Earth's orbit.

During a military trade show in Russia in August 2021, KB Arsenal circulated a leaflet with logos of the company and its parent Roskosmos State Corporation and featuring more renderings of what was identified as the Zevs Orbital Complex. The accompanying text said that the vehicle was being developed for launch into a 1,000-km (621 mi) (circular?) orbit on an Angara-A5 rocket from Vostochny spaceport. The space tug's mission was described as delivering scientific equipment toward the Moon and the planets of the Solar System.

In May 2022, Head of Roskosmos, Dmitry Rogozin admitted that the Zevs program had lacked funding, apparently reflecting new realities after Russia's escalation of the war against Ukraine on February 24, 2022. At the same time, Rogozin published photos of a thermal and vacuum chamber at the GNTs Keldysh Center in Moscow built for testing operation of high-speed turbines converting heat energy into electric power. It is a critical mechanism for the nuclear electric space tug, because it converts heat energy generated by the nuclear reactor into electricity required in huge quantities for the operation of high-thrust plasma engines.

The experimental stand at Keldysh operated in conjunction with a heat-generating unit with an output of 2 Megawatts, which simulates thermal loads produced by the reactor. According to Rogozin, during the successful tests of the experimental unit on May 13, 2022 the temperature of the driving substance at the turbine's entrance had reached 1,200° K (1,700°F) and the turbine's speed had reached 34,000 rotations per minute. The subsequent tests aim to spin the turbine up to a target speed of 60,000 rotations per minute or 1,000 rotations per second, Rogozin said.

According to the Keldysh center, its multi-functional stand could accommodate experimental gas turbines and their components with a power of up to 250 Kilowatts. The center also had a cryogenic and vacuum test facility for test firing of Hall and ion engines with a power of up to 35 Kilowatts.

Rogozin also described a joint Russo-Belorussian effort to develop turbine blades which could operate at temperatures of 1,500° K (2,240°F) and higher. Specialists reportedly tried several candidate materials from metal alloys to ceramics and composite materials. The development of new blades with higher heat resistance could allow higher temperature in the turbine, which, in turn would make it possible to reduce the size and weight of a radiator system for rejecting excessive heat into space aboard the space tug.

In the aftermath of the visit by Roskosmos head Yuri Borisov to GNTs Keldysh in Moscow on December 1, 2022, the State Corporation released a photo clearly showing the turbine section of the Zevs space tug on a movable tray of the vacuum facility indicating still ongoing tests. According to the known configuration of the vehicle, its power-producing nuclear reactor would be located next to the visible truss structure, but the official photo was probably intentionally framed to avoid revealing whether the prototype of the power source was a part of the test."

Specifications

The TEM is constructed by the Russian Keldysh Research Center, NIKIET (Research and Design Institute of Power Engineering) Institute, and Rosatom.

The TEM project was started in 2009. The Zeus mission was proposed by Roscosmos and the Russian Academy of Sciences in May, 2021.

The first reactor tests are scheduled for the early 2020s. The first orbital flight test of the reactor is planned for no earlier than 2030. The first mission, named Zeus, is envisioned to operate for 50 months and deliver payloads to the Moon, Venus, and Jupiter through multiple gravity assists. The specifications include:

- Reactor
  Coolant: 78% helium/22% xenon.
  Heat power: 3.8 MW
  Electric power: 1 MW
- Spacecraft
  Mass: 20,290 kg (49,140 lbs); limited by Angara 5 carrying capacity.
  Thrust: 18 N (13.28 ft-lbs)
  Specific impulse: 7000 s
- Angara Launch Vehicle

IMAGE LINKS

Fig. A6.1 RD-410 Nuclear Thermal Rocket
http://www.astronautix.com/nails/r/rd0410.jpg

Fig. A6.2 TEM Zeus in orbit
https://www.russianspaceweb.com/images/rockets/upper_stages/tem/arsenal/ar miya2021_orbit_1.jpg

Fig. A6.3 TEM components
https://www.russianspaceweb.com/images/rockets/upper_stages/tem/arsenal/te m_components_1.jpg

Fig. A6.4 The TEM folded, assembled and explained
https://www.russianspaceweb.com/images/rockets/upper_stages/tem/arsenal/te m_folded_1.jpg

Appendix 7

Chinese Nuclear Rocket Program

A7.1 BACKGROUND

Today, China has one of the most active space programs in the world. It conducts either highest or second highest number of orbital launches each year. It is operating a satellite fleet consisting of large number of communication, navigation, remote sensing and scientific research satellites. With spacecraft reaching as far as the Moon and Mars, China has conducted multiple complex extraterrestrial exploration missions, including landing or even sample-return. In the near future, the Chinese space program is steadily pursuing a manned mission to the Moon, space transportation, in-orbit maintenance of spacecraft, space telescope, counter-space capabilities, quantum communications, orbiter and sample-return missions to Mars, and exploration missions throughout the Solar System and deep space.

A7.1 Brief Timeline

- On October 15, 2003, astronaut Yang Liwei was put into space aboard Shenzhou 5 spacecraft by a Long March 2F rocket for more than 21 hours. China became the third country capable of conducting independent human spaceflight.

- Around the same time, China began the preparation of extraterrestrial exploration, starting with the Moon. The Chinese Moon orbiting program was approved in January 2004 and was later transformed into Chinese Lunar Exploration Program. The first lunar orbiter Chang'e 1 was successfully launched on October 24, 2007, and was inserted into

Moon orbit on November 7, making China the fifth nation to successfully orbit the Moon.

- In March 2008, the Chinese National Space Agency (CNSA), along with the Commission for Science, Technology and Industry for National Defense, was merged into the newly formed Ministry of Industry and Information Technology.

- On September 27, 2008, two crew members of the Shenzhou 7 carried out China's first EVA.

- On September 29, 2011, China launched Tiangong-1, the first prototype of Chinese space station module. The following Shenzhou 8, Shenzhou 9 and Shenzhou 10 missions proved that China had developed critical human spaceflight capabilities like space docking and berthing.

- China began its first interplanetary exploration attempt in 2011 by sending Yinghuo-1, a Mars orbiter, in a joint mission with Russia. Yet it failed to leave Earth orbit due to the failure of Russian launch vehicle.

- China then turned its focus back to the Moon by attempting the challenging lunar soft landing. On December 14, 2013, China successfully landed Chang'e 3 Moon lander and its rover Yutu on the Moon surface. It made China the third country in the world capable of performing lunar soft landing, just after USSR and the United States.

- In 2016, Tiangong-2 and Shenzhou 11 were launched into Low Earth orbit. A 33-day crewed spaceflight mission proved that China was ready for a long-term space station built and maintained by its own.

- In 2018, China performed more orbital launches than any other country on the planet for the first time in history.

- On January 3, 2019, Chang'e 4 conducted the first-ever soft landing on the far side of the Moon by any country, followed by 2020's Chang'e 5, a complex and successful lunar sample return mission, marking the completion of the three goals (orbiting, landing, returning) of the first stage of the lunar exploration program.

- On June 23, 2020, the final satellite of Beidou was successfully launched by a Long March 3B rocket. On July 31, 2020, Chinese leader Xi Jinping formally announced the commissioning of BeiDou Navigation Satellite System.

- On April 29, 2021, Tianhe, the 22-tonne core module of Tiangong space station, was successfully launched into Low Earth orbit by a Long March 5B rocket, indicating the beginning of the construction of the Chinese Space Station.

- Ever since the failure of Yinghuo-1, the Chinese space agency had embarked on its independent Mars mission. On July 23, 2020, China launched Tianwen-1, which included an orbiter, a lander, and a rover, on a Long March 5 rocket to Mars. The Tianwen-1 was inserted into Mars orbit in February 2021 after a six-month journey, followed by a successful soft landing of the lander and Zhurong rover on May 14, 2021, making China the third nation to both land softly on and establish communication from the Martian surface, after the Soviet Union and the United States.

- On April 24, 2022, a rocket was launched on high altitude zero-pressure helium balloon from Lenghu in the northwest China's Qinghai Province, which saves fuel and reduces overall costs.

A7.2 CHINA'S SPACE PROGRAM GOES NUCLEAR

In November, 2021, the South China Morning Post announced that China's space program will go nuclear to power future missions to the moon and Mars. A prototype design for a powerful nuclear reactor had been completed and some components have been built, according to researchers. The reactor can generate one megawatt of electric power; a 100 times more powerful than a similar device NASA plans to put on the surface of the moon by 2030. The project was launched with funding from the central government in 2019. Although technical details and the launch date were not revealed, the engineering design of a prototype machine was completed recently and some critical components have been built.

The only publicly known nuclear device it has sent into space is a tiny radioactive battery on Yutu 2, the first rover to land on the far side of the Moon in 2019. That device could only generate a few watts of heat to help the rover during long lunar nights. China returned the first lunar samples in nearly 50 years from the Moon.

For a 5:06 minute on the history of the Chinese Space Program, go to: https://youtu.be/vUR60ECbf8A

A good example of the state of China's space program is shown in the following video. For a 2:30 minute video of China's Long March 5B vehicle, go to: https://youtu.be/BVeGH1W_Mfw

China has identified space nuclear propulsion as a key element of its plan to become the pre-eminent space power by mid-century, specifically identifying the technology as a way to access and exploit space resources.

A senior Chinese space exploration official, Wu Weiren, director of the newly-established Tiandu Deep Space Exploration Laboratory, in 2019 called for breakthroughs in nuclear power for space, to meet future mission requirements. "We are now developing a new system that uses nuclear energy to address the Moon station's long-term, high-power energy demands," said Wu Weiren, chief designer of the country's Lunar Exploration Program. China is also expecting to send astronauts to the moon within the next decade, "The station will comprise a lander, a hopper, an orbiter, and a rover that can carry astronauts around. The station could also be a launch pad for further missions in space and includes a communication facility that allows the outpost to send signals to Earth and other planets such as Mars." Chang'e 7 and 8, two robotic missions scheduled to launch in 2026 and later, will explore the south pole and lay the foundation for the outpost, Wu added.

Chinese civilian mission proposals using a nuclear space reactor to provide power for propulsion include Voyager-like missions which would see a pair of spacecraft towards the nose and tail of the heliosphere and potentially a third perpendicular to the plane of the ecliptic. A Neptune orbiter mission concept using electric propulsion powered by a nuclear reactor has also been published.

China has been expanding its space transportation and deep space capabilities in recent years, successfully developing cryogenic rockets to facilitate lunar, Mars and space station projects. It is now working on reusable launchers, super heavy-lift rockets and a two-stage-to-orbit reusable spaceplane system.

Last year, a researcher with the Chinese Academy of Sciences described nuclear power as "the most hopeful solution," as chemical fuel and solar panels are no longer sufficient to meet the energy demands of the space exploration missions, the South China Morning Post reported. According to the report, China is developing a nuclear reactor that is expected to generate one megawatt of electric power, enough to power around a thousand households. The plan puts China in direct competition with the U.S., which is aiming to set up a similar outpost over the next decade.

The Planetary Society's view is that it is hard to gauge China's progress in nuclear power in space, in part due to the sensitive nature of nuclear technology. It is clear, though, that researchers in China are, as with so many areas of space activity, assessing international progress and possibilities in this area. More practically, there are already proposals for reactors in Chinese space missions, including the uranium-powered, miniature integrated nuclear reactor design with gravity independent autonomous circulation (ACMIR).

Further signs that nuclear power is part of China's future plans for space exploration can also be seen in the fact that the country is considering adding a third spacecraft to its planned mission to study the nose and tail of the heliosphere. The additional probe would, if selected, head away from the Sun perpendicular to the plane of the ecliptic and be powered by a nuclear reactor.

Tom Colvin, a researcher at the Institute for Defense Analysis (IDA)'s Science and Technology Policy Institute, commenting on the uses of nuclear fission for space, noted that nuclear fission reactors are uniquely suitable for enabling power and propulsion for robotic missions to the outer planets, but also power for crewed missions to the surface of Mars. Developing a fission reactor for these missions would involve "myriad technical challenges," but he notes there are no showstoppers. "It simply takes time and money."

Russia has used over 30 fission reactors in space, according to the World Nuclear Association, while the U.S. has flown just one: SNAP-10A (System for Nuclear Auxiliary Power) in 1965. However, NASA is currently working on fission systems to power lunar surface missions under the Artemis umbrella. See Chapter 3.

References

Ceramic composite moderators as replacements for graphite in high temperature micro reactors.
Bin Cheng a, Edward M. Duchnowski b, David J. Sprouster a, Lance L. Snead a, Nicholas R. Brown b, Jason R. Trelewicz , 2022

Nuclear Thermal Propulsion
Mark D. DeHart, Sebastian Schunert and Vincent M. Labouré
Reviewed: February 23, 2022, Published: April 28, 2022

Experimental evaluation on heat transfer limits of sodium heat pipe with screen mesh for nuclear reactor system
Zhixing Tian, Jiarui Zhang, Chenglong Wang, Kailun Guo, Yu Liu, Dalin Zhang, Wenxi Tian, Suizheng Qiu, G.H. Su, 2022
Shaanxi Key Lab. of Advanced Nuclear Energy and Technology, Jiaotong University, Xi'an , China, Science and Technology on Reactor System Design Technology Laboratory, Nuclear Power Institute of China, Chengdu, China

Understanding TRISO Coated Particle Neutron Irradiation Behavior:Evolution of Advanced MicroAnalysis and Electron Microscopy Approaches,
Isabella J Van Rooyen, Idaho National Laboratory, Contract DE-AC07-05ID14517, 2021

Advance in and prospect of moderator materials for space nuclear reactors
Zhihui Wang, Fangchen Liu, Zhancheng Guo, Jiandong Zhang, Lijun Wang, Guoqing Yan, National Natural Foundation of China, January 2021

Space Nuclear Propulsion for Human Mars Exploration. Washington, DC: National Academies of Sciences, Engineering, and Medicine.
The National Academies, 2021

Space Nuclear Propulsion for Human Mars Exploration

Contributors: National Academies of Sciences, Engineering, and Medicine; Division on Engineering and Physical Sciences; Aeronautics and Space Engineering Board; Space Nuclear Propulsion Technologies Committee National Academies of Sciences, Engineering, and Medicine. The National Academies Press, 2021

Crewed Mars Mission Mode Options for Nuclear Electric/Chemical Hybrid Transportation Systems. Patrick R. Chai, Katherine T. McBrayer, Andrew C. McCrea, Raymond G. Merrill, and Akshay Prasad, Min Qu‖
AIAA 2021-4136, Session: Mission Architectures – Mars, 2021

Nuclear Thermal Propulsion Dynamic Modeling with Modelica
by Jordan D Rader, Michael B Smith, Michael S Greenwood, Thomas J Harrison. Nuclear and Emerging Technologies for Space (NETS), 2019

Overview of NASA's Solar Electric Propulsion Project
IEPC-2019-836
Presented at the 36th International Electric Propulsion Conference
University of Vienna, Austria, September 15 – 20, 2019

Kilopower Reactor Using Stirling Technology (KRUSTY) Nuclear Ground Test Results and Lessons Learned, NASA/TM—2018-219941
Marc A. Gibson, Glenn Research Center, David I. Poston and Patrick McClure, Consolidated Nuclear Security, LLC, Los Alamos National Laboratory, Thomas J. Godfroy, Vantage Partners, LLC, Huntsville, Alabama Maxwell H. Briggs, Glenn Research Center, James L. Sanzi, Vantage Partners, LLC, 2018

Development of High Power Hall Thruster Systems to Enable the NASA Exploration Vision.
Space Propulsion 2018
Jerry Jackson, Joseph Cassady, May Allen, Roger Myers, Todd Tofil, Dan Herman and Eric Pencil. Seville, Spain, May 14-18, 2018

Additive Manufactured Pressure Vessel Shell
W. Tam, Kamil Wlodarczyk, G. Kawahara
Space Propulsion, 2018, Seville, Spain; May 14-18, 2018

Nuclear Engineering and Design
Volume 340, 15 December 2018, Pages 9-16
A miniature integrated nuclear reactor design with gravity independent autonomous circulation
Qin Zhou a, Yan Xia b c, Guoqing Liu b c, Xiaoping Ouyang, 2018

Lifecycle of the ESS Moderator and Reflector System
M Kickulies, Y Beßler2, Y Lee and D Lyngh. Published under license by IOP Publishing Ltd. Journal of Physics: Conference Series, Volume 1021, 22nd meeting of the International Collaboration on Advanced Neutron Sources (ICANS XXII) March 27-31,2017, Oxford, United Kingdom

Nuclear Cryogenic Propulsion Stage (NCPS) Fuel Element Testing in the Nuclear Thermal Rocket Element Environmental Simulator (NTREES)
Author, Emrich, William J., Jr.
NASA Marshall Space Flight, July 10, 2017
NASA/TM—2016-219393
Atomic Power in Space II: A history of space nuclear power and propulsion in the United States. INL/EXT-15-34409
Prepared by Idaho National Laboratory, Battelle Energy Alliance, LLC
Space Nuclear Power and Isotope Technologies Division, September 2015

Reflector and Control Drum Design for a Nuclear Thermal Rocket
Tyler Goode, Jeffrey Clemens, Michael Eades, J. Boise Pearson
Nuclear and Emerging Technologies for Space 2015 (NETS-2015)

Heat Pipe for Aerospace Applications—An Overview
K. N. Shukla, PRERANA CGHS Ltd., Gurgaon, India published March 26, 2015

Conventional and Bimodal Nuclear Thermal Rocket (NTR) Artificial Gravity Mars Transfer Vehicle Concepts
Stanley K. Borowski, David R. McCurdy and Thomas W. Packard, Vantage Partners, LLC. AIAA 2014-3623 Session: Nuclear Thermal Propulsion II - Missions, Vehicles & Architectures, July 25, 2014

A One-year Round Trip Crewed Mission to Mars using Bimodal Nuclear Thermal and Electric Propulsion (BNTEP)
Laura M. Burke, Stanley K. Borowski, David R. McCurdy and Thomas Packard. AIAA 2013-4076, Session: Nuclear Thermal Propulsion III: Missions, Vehicles and Architectures, July 12, 2013

A review of the Los Alamos effort in the development of nuclear rocket propulsion.
Franklin Durham, William Kirk, and Richard Bohl.
AIM-91-3449 Los Alamos National Laboratory, August 17, 2012.

The Nuclear Cryogenic Propulsion Stage
Michael G. Houts, Tony Kim, William J. Emrich, Robert R. Hickman, Jeramie W. Broadway, Harold P. Gerrish, Glen Doughty, Anthony Belvin, Stanley K. Borowski, and John Scott, 2012

Fabrication and Testing of CERMET Fuel Materials for Nuclear Thermal Propulsion
Robert R. Hickman, Jeramie W. Broadway, and Omar R. Mireles
NASA Marshall Space Flight Center, Huntsville, AL, 2012

An Overview of Facilities and Capabilities to Support the Development of Nuclear Thermal Propulsion
Nuclear and Emerging Technologies for Space 2011
James Werner Sam Bhattacharyya Mike Houts. February 2011

Ongoing Space Nuclear Systems Development in the United States
Michael Houts, Shannon Bragg-Sitton. Published 2011

An Overview of Facilities and Capabilities to Support the Development of Nuclear Thermal Propulsion
Nuclear and Emerging Technologies for Space 2011
James Werner, Sam Bhattacharyya, Mike Houts, February 2011

Ongoing Space Nuclear Systems Development in the United States
Shannon M. Bragg-Sitton, James E. Werner, Stephen G. Johnson, Michael G. Houts, Donald T. Palac, Lee S. Mason, David I. Poston, and A. Lou Qualls.
Space Nuclear Systems and Technology Division
Idaho National Laboratory, 2011

Fabrication of uranium oxycarbide kernels and compacts for HTR fuel
Jeffrey A. Phillips, Scott G. Nagley, Eric L. Shaber. Idaho National Laboratory, Babcock and Wilcox Nuclear Operations Group, Available online November 14, 2011.

Radioisotope electric propulsion (REP): A near-term approach to nuclear propulsion
George R. Schmid, David H.Manzell, Hani Kamhawia,Tibor Kremica, Steven R.Olesona, John W.Dankanich, Leonard A.Dudzinskic.
NASA Glenn Research Center, Gray Research, NASA Headquarters, 2009

Nuclear Thermal Rocket Element Environmental Simulator (NTREES) Upgrade Activities. William J Emrich, Jr, Robert P. Moran, and J. Boise Pearson NASA Marshall Space Flight Center, 2008

To the End of the Solar System- The Story of the Nuclear Rocket, Dewar, J.A., "Second Edition, Apogee Books Space Series, 2007 Available on Amazon

Specular reflection of thermal neutrons from Gd-containing layers and optimization of antireflective under layers for polarizing coatings

N.K. Pleshanov, B.G. Peskov, V.M. Pusenkov, V.G. Syromyatnikov, and A.F. Schebetov. Petersburg Nuclear Physics Institute, St. Petersburg, Russia 24 January 2006.

The Role of Nuclear Power and Nuclear Propulsion in the Peaceful Exploration of Space.
International Atomic Energy Agency, Vienna, 2005

Optimal Moderation in the Pebble-Bed Reactor for Enhanced Passive Safety and Improved Fuel Utilization
Abderrafi M. Ougouag, Hans D. Gougar, William K. Terry, Ramatsemela Mphahlele, and Kostadin N. Ivanov. Idaho National Engineering and Environmental Laboratory, 2004

"Bimodal" Nuclear Thermal Rocket (BNTR) Propulsion for Future Human Mars Exploration Missions
Dr. Stanley K. Borowski
Space Transportation Office NASA Glenn Research Center, November 5-6, 2003

Assessment of the facilities on Jackass Flats and other Nevada test site facilities for the new nuclear rocket program
AIP Conference Proceedings 271,
George Chandler, Donald Collins, Ken Dye, Craig Eberhart, Michael Hynes, Richard Kovach, Robert Ortiz, Jake Perea, and Donald Sherman, 1993

Nuclear propulsion: A vital technology for the exploration of Mars and the planets beyond. Borowski, Stanley K. NASA Lewis Research Center, NASA Technical Memorandum (TM) 101354. January 1, 1988

Glossary

A

ACT	Advanced Cooling Technologies, Inc.
AEC	Atomic Energy Commission
ALSEP	Apollo Lunar Science Experiment Package
AMRC	Advanced Manufacturing Research Centre
ARC	Advanced Research Concepts
ARDP	Advanced Reactor Demonstration Program

B

BEA	Battelle Energy Alliance LLC
BWXT	BWX Technologies

C

CAD	Computer Aided Design
CAM	Computer Aided Manufacturing
CASL	Consortium for Advanced Simulation of Light Water Reactors
CASC	China Aerospace Science and Technology Corporation
CCHPs	Constant Conductance Heat Pipes
CERMET	Ceramic/ Metal fuel
CFEET	Compact Fuel Element Environmental Tester
CFD	Computational Fluid Dynamics
CNSA	Chinese National Space Agency
COPUOS	Committee on the Peaceful Uses of Outer Space
CSNR	Center Space Nuclear Research

D

DAF	Device Assembly Facility
DARPA	Defense Advanced Research Projects Agency
DART	Double Asteroid Redirection Test
DOE	Department of Energy
DRACO	Demonstration Rocket for Agile Cislunar Operations

| DSS | Deployable Space Systems Inc. |
| DUFF | Demonstration Using Flattop Fissions |

E

ETTP	East Tennessee Technology Park
E-MAD	Engine Maintenance, Assembly, and Disassembly
EL3	European Large Logistics Lander
EBOR	Experimental Beryllium Oxide Reactor

F

FAMU-FSU	Florida A&M University and Florida State University
FCM®	Fully Ceramic Micro-encapsulated
FEA	Finite Element Analysis/Method
FCM®	Fully Ceramic Micro-encapsulated

G

| GCD | Game Changing Development (Program) |
| GPHS | General Purpose Heat Source |

H

HALEU	High-Assay, Low-Enriched Uranium
HEU	Highly Enriched Uranium
HRP	Human Research Program
HRS	Heat Rejection System
HTWG	Heat Transfer Working Group

I

IAEA	International Atomic Energy Agency
INL	Idaho National Laboratory
ISRO	Indian Space Research Organisation
ISTV	In-Space Transportation Vehicles
ISRU	In-Situ Resource Utilization
ISS	International Space Station
ITER	International Thermonuclear Experimental Reactor

J

JET   Joint European Torus
JH-APL   Johns Hopkins Applied Physics Laboratory
JWST   James Webb Space Telescope

K

KAMINI   Kalpakkam Mini Reactor
Kgf   Kilogram force
kPa   Kilopascal (pressure unit)

L

LANL   Los Alamos National Laboratory
LASL   Los Alamos Scientific Laboratory
LEU-NTP   Low-Enriched Uranium Nuclear Thermal Propulsion

M

MCNP   Monte Carlo N-Particle Transport
MGCR   Maritime Gas-Cooled Reactor
MMR®   Micro Modular Reactor
MMRTG   Multi-Mission Radioisotope Thermoelectric Generator
MSFC   Marshall Space Flight Center
M&S   Modeling and Simulation
MURR   University of Missouri-Columbia Research Reactor

N

NSPM   National Security Presidential Memorandum
NCPS   Nuclear Cryogenic Propulsion Stage
NEP   Nuclear Electric Propulsion
NERVA   Nuclear Engine for Rocket Vehicle Application
NNSA   National Nuclear Security Administration
NPSS   Numerical Propulsion System Simulation
NRC   Nuclear Regulatory Commission
NRDS   Nuclear Rocket Development Station

NRX	NERVA Reactor Experiment
NTR	Nuclear Thermal Rocket
NTREES	Nuclear Thermal Rocket Element Environment Simulator

O

ONRAMP	ORNL's Nuclear Resources Analysis and Modeling Portfolio
ORNL	Oak Ridge National Laboratory

P

PADME	Power-Adjusted Demonstration Mars Engine
PCS	Power Conversion System
PHS	Public Health Service
PIE	Post-Irradiation Examination (facility)
PMAD	Power Management and Distribution
PPU	Power Processing Unit

R

REP	Radioisotope Electric Propulsion
RHU	Radioisotope Heating Units
RIFT	Reactor In-Flight Tests
R-MAD	Reactor Maintenance, Assembly, and Disassembly
RNSD	Reactor and Nuclear Systems Division
RPS	Radioisotope Power Systems
RTG	Radioisotope Thermoelectric Generator

S

SCCTE	Space Capable Cryogenic Thermal Engine
SEP	Solar Electric Propulsion
SHARE	Space Station Heat Pipe Radiator Element
SNAP	Systems Nuclear Auxiliary Power
SNPO	Space Nuclear Propulsion Office (MSFC)
SNPP	Space Nuclear Power and Propulsion

SSTO	Single-Stage-To-Orbit
STG	Space Task Group
STMD	Space Technology Mission Directorate (NASA)
STS	Space Transportation System
SWET	Simulated Water Entry Test
SwRI	Southwest Research Institute

T

TAMUS	Texas A&M University System
TEM	Transport and Energy Module
TRANSFORM	Transient Simulation Framework of Reconfigurable Models
TREAT	Transient Reactor Test (INL)
TRISO	Tri-Structural IsOtropic
TRL	Technology Readiness Level

U

UAH	University of Alabama Huntsville
UC	University of California
UK	United Kingdom
UKSA	United Kingdom Space Agency
UNOOSA	United Nations Office for Outer Space Affairs
USNC-Tech	Ultra Safe Nuclear Corporation Technologies

V

VCHP	Variable Conductance Heat Pipes

W, X, Y

Z

ZIRCEX	Hybrid Zirconium Extraction Process

Terminology

Critical Mass

Although two to three neutrons are produced for every fission, not all of these neutrons are available for continuing the fission reaction. If the conditions are such that the neutrons are lost at a faster rate than they are formed by fission; the chain reaction will not be self-sustaining. At the point where the chain reaction can become self-sustaining; this is referred to as critical mass.

In an atomic bomb, a mass of fissile material greater than the critical mass must be assembled instantaneously and held together for about a millionth of a second to permit the chain reaction to propagate before the bomb explodes. The amount of a fissionable material's critical mass depends on several factors; the shape of the material, its composition and density, and the level of purity.

A sphere has the minimum possible surface area for a given mass, and hence minimizes the leakage of neutrons. By surrounding the fissionable material with a suitable neutron "reflector", the loss of neutrons can be reduced and the critical mass can be reduced. By using a neutron reflector, only about 11 pounds (5 kilograms) of nearly pure or weapon's grade plutonium 239 or about 33 pounds (15 kilograms) uranium 235 is needed to achieve critical mass.

Controlled Nuclear Fission

To maintain a sustained controlled nuclear reaction, for every 2 or 3 neutrons released, only one must be allowed to strike another uranium nucleus. If this ratio is less than one, then the reaction will die out; if it is greater than one it will grow uncontrolled (an atomic explosion). A neutron absorbing element must be present to control the amount of free neutrons in the reaction space. Most reactors are controlled by means of control rods that are made of a strongly neutron-absorbent material such as boron or cadmium.

Controlled Nuclear Chain Reaction

In addition for the need to capture neutrons, the neutrons often have too much kinetic energy. These fast neutrons are slowed through the use of a moderator such as heavy water and ordinary water. Some reactors use graphite as a moderator, but this design has several problems. Once the fast neutrons have been slowed, they are more likely to produce further nuclear fissions or be absorbed by the control rod.

Hodoscope

A hodoscope is an instrument used in particle detectors to detect passing charged particles and determine their trajectories; for example for observing the paths of subatomic particles, especially those arising from cosmic rays.

Hodoscopes are characterized by being made up of many segments; the combination of which segments record a detection and is then used to infer where the particle passed through hodoscope.

Specific Impulse (Isp)

The thrust produced per unit rate of consumption of the propellant. It is usually expressed in pounds of thrust per pound of propellant used per second; which is a measure of the efficiency of a rocket engine. Also the change in momentum per unit mass of propellant.

Note that a nuclear thermal propulsion rocket will achieve twice the efficiency of a chemical rocket; or twice the Isp of a chemical rocket or about 900 sec.

High Assay Low Enriched Uranium (HALEU)

By definition, High Assay Low-Enriched Uranium (HALEU) is enriched between 5% and 20% and is required for smaller U.S. advanced reactors designs that get more power per unit of volume. HALEU will also allow developers to optimize their systems for longer life cores, increased efficiencies and better fuel utilization.

The DOE projects that more than 40 metric tons of HALEU will be needed before the end of the decade, with additional amounts required each year, to deploy a new fleet of advanced reactors. To help mitigate that risk, DOE is exploring three options to support the testing and demonstration of these advanced reactors with HALEU fuel.

TRISO

The most critical element in the design of advanced nuclear reactors is a robust fuel that can withstand very high temperatures without melting. Nuclear thermal reactors will likely use tri-structural isotropic (TRISO) particle fuel, developed and improved over 60 years. BWTX has its own proprietary version (TRISO-X) to ensure supply and quality control.

About the Author

Manfred "Dutch" von Ehrenfried had the very good fortune to have interviewed with the NASA Space Task Group (STG) at the Langley Research Center the day before Alan Shepard was launched on MR-3. At the time, he had very little knowledge of Project Mercury and thought that since his degree was in physics, he would be working in that area. As fate would have it, he was assigned to the Flight Control Operations Section under Gene Kranz who became his supervisor and mentor. Most of his work for Project Mercury would be in the areas of mission rules, countdowns, operational procedures and coordination with the remote tracking station flight controllers. During his first six months, he was in training to be a flight controller and spent Mercury Atlas-4 and 5 at the NASA Goddard Space Flight Center learning communications between the Mercury Control Center and the Manned Space Flight Network which included all the world-wide tracking stations.

His first mission as a flight controller was in the Mercury Control Center for John Glenn's flight on Mercury Atlas-6 learning the Operations and Procedures flight control position under Gene Kranz. He supported the remaining manned orbital Mercury missions of Carpenter, Schirra and Cooper in the Mercury Control Center as well as the Gemini simulations and the Gemini 3 mission of John Young and Gus Grissom.

After the move of the Space Task Group from Langley to Houston, Dutch supported the manned Gemini missions and was Assistant Flight Director for Gemini 4-7 which including the first EVA by Ed White and the first rendezvous in space on GT-6 and 7. In 1966, he became a Guidance Officer on Apollo 1 and after the accident and stand-down, became the Mission Staff Engineer on Apollo 7 and backup on Apollo 8. During this period, Dutch was also an Apollo Pressure Suit Test Subject which afforded him the opportunity to test pressure suits in the vacuum chamber to over 400,000 feet including one test of Neil Armstrong's suit. He also experienced 9 g's in the centrifuge and flew in the zero-g aircraft. He had his own Apollo A7LB Skylab suit.

These experiences afforded him the opportunity to join the Earth Resources Aircraft Program. He was the first sensor equipment operator and Mission Manager on the high altitude RB-57F. His experience as the sensor operator provided him experience working with scientists to operationally achieve their research objectives. These flights required wearing a full pressure suit as they generally flew at altitudes in the range 65,000 to 67,000 ft; one flight actually achieved 70,000 ft to photograph the Tago-Sato-Kosaka comet.

During 1970 and 1971, the author was the Chief of the Science Requirements and Operations Branch at NASA JSC. This Branch was responsible for the definition, coordination and documentation of science experiments assigned to Apollo and Skylab. This included the Apollo Lunar Surface Experiment Packages (ALSEP) left on the Moon and the experiments in lunar and Earth orbit. The ALSEP packages included seismic sensors, magnetometers, spectrometers, ion detectors, heat flow sensors, charged particle and cosmic ray detectors, gravity measurements and more. The lunar orbit experiments included the Scientific Instrument Module (SIM) Bay cameras and sensors and the Particle and Fields subsatellites launched prior to leaving lunar orbit. The work also defined the astronauts' procedures for deploying the packages and conducting experiments on the Moon and in lunar orbit. He also spent one year with a contractor at NASA Goddard on the Earth Resources Technology Satellite (ERTS); later called Landsat 1.

Dutch also worked in the nuclear industry for seven years evaluating fuel cycle facilities and reactors applying the concept of NASA mission rules to the facilities' reactions to various situations and potential terrorist threats. When with the Nuclear Regulatory Commission, he was Chief of the Test and Evaluation Branch which afforded him the opportunity to evaluate both low and high enriched uranium facilities, the Barnwell Plutonium Reprocessing Facility as well as reactors and spent fuel facilities. This experience lead him to write the book; *Nuclear Terrorism-A Primer.*

As a contractor, he also worked with the original NASA Headquarters Space Station Task Fork for ten years. He has written several books about his experiences. They can be seen by going to https://author.amazon.com/books or by going to his website: www.dutch-von-ehrenfried.com. For the past 25 years, he has been a working in the finance and insurance fields.

Fig. AA.1 The author in late 1961 as a young NASA Space Task Group Flight Controller. Center left: The author at the console to the left of Gene Kranz and George Low. Center right: The author testing Neil Armstrong's suit to an equivalent altitude of 400,000 ft in the vacuum chamber at the Manned Spacecraft Center. Bottom left: The author getting a pre-flight check. Lower right, wearing the Air Force A/P22S-6 full pressure suit required for the all high altitude RB-57F flights. All photos courtesy of NASA.

Index

A

B

H

I

T

Made in the USA
Las Vegas, NV
15 August 2023

76103756R00162